Dr. T.

MW00335182

SIGNAL THEORY METHODS IN MULTISPECTRAL REMOTE SENSING

WILEY SERIES IN REMOTE SENSING

Jin Au Kong, Editor

SIGNAL THEORY METHODS IN MULTISPECTRAL REMOTE SENSING

David A. Landgrebe

School of Electrical & Computer Engineering
Purdue University
West Lafayette, IN 47097-1285

WILEY-INTERSCIENCE

A JOHN WILEY & SONS PUBLICATION

Copyright © 2003 by John Wiley & Sons, Inc. All rights reserved.

Published by John Wiley & Sons, Inc., Hoboken, New Jersey.
Published simultaneously in Canada.

No part of this publication may be reproduced, stored in a retrieval system or transmitted in any form or by any means, electronic, mechanical, photocopying, recording, scanning or otherwise, except as permitted under Section 107 or 108 of the 1976 United States Copyright Act, without either the prior written permission of the Publisher, or authorization through payment of the appropriate per-copy fee to the Copyright Clearance Center, Inc., 222 Rosewood Drive, Danvers, MA 01923, (978) 750-8400, fax (978) 750-4744, or on the web at www.copyright.com. Requests to the Publisher for permission should be addressed to the Permissions Department, John Wiley & Sons, Inc., 111 River Street, Hoboken, NJ 07030, (201) 748-6011, fax (201) 748-6008, e-mail: permreq@wiley.com.

Limit of Liability/Disclaimer of Warranty: While the publisher and author have used their best efforts in preparing this book, they make no representation or warranties with respect to the accuracy or completeness of the contents of this book and specifically disclaim any implied warranties of merchantability or fitness for a particular purpose. No warranty may be created or extended by sales representatives or written sales materials. The advice and strategies contained herein may not be suitable for your situation. You should consult with a professional where appropriate. Neither the publisher nor author shall be liable for any loss of profit or any other commercial damages, including but not limited to special, incidental, consequential, or other damages.

For general information on our other products and services please contact our Customer Care Department within the U.S. at 877-762-2974, outside the U.S. at 317-572-3993 or fax 317-572-4002.

Wiley also publishes its books in a variety of electronic formats. Some content that appears in print, however, may not be available in electronic format.

Library of Congress Cataloging-in-Publication Data is available.

ISBN 0-471-42028-X

Printed in the United States of America.

10 9 8 7 6 5 4 3 2 1

Contents

Preface

This book is, in some ways, a sequel to the book *Remote Sensing: The Quantitative Approach* (McGraw-Hill, 1978). RSQA was written by key staff members of Purdue's Laboratory for Applications of Remote Sensing and was one of the first textbooks on multispectral remote sensing. It has been out of print for several years.

The general concept for the present book is to focus on the fundamentals of the analysis of multispectral and hyperspectral image data from the point of view of signal processing engineering. In this sense the concept is unique among currently available books on remote sensing. Rather than being a survey in any sense, it is to be a textbook focusing on how to analyze multispectral and hyperspectral data optimally. Indeed, many topics common in the literature are not covered, because they are not really germane or required for that purpose. The end goal is to prepare the student/reader to analyze such data in an optimal fashion. This is especially significant in the case of hyperspectral data

The book consists of three parts:

1. *Introduction.* A single chapter intended to give the reader the broad outline of pattern recognition methods as applied to the analysis of multivariate remotely sensed data. Upon the completion of study of this chapter, the reader should have the broad outline of what may be learned from the rest of the book, and what the special strengths of the multivariate multispectral approach are.

2. *The basic fundamentals.* This consists of two rather long chapters. The first deals with the scene and sensor parts of a passive, optical remote sensing system. The second deals with basic pattern recognition. This second part covers the statistical approach in detail,

including first and second order decision boundaries, error estimation, feature selection, and clustering. It concludes with an initial integrated look at how one actually analyzes a conventional multispectral data set.

3. *In-depth treatments and the "bells and whistles."* This part goes into more depth on the basics of data analysis and covers additional methods necessary for hyperspectral analysis, including an array of examples, spectral feature design, the incorporation of spatial variations, noise in remote sensing systems, and requirements and methods for preprocessing data.

The material covered has been developed based on a 35 year research program. This work was associated with an early airborne system, such space systems as the Landsat satellite program, and later satellite and aircraft programs. It has grown out of the teaching of a senior/graduate level course for more than a decade to students from electrical and computer engineering, civil engineering, agronomy, forestry, agricultural engineering, Earth and atmospheric sciences, mathematics, and computer science. The text should be of interest to students and professionals interested in using current and future aircraft and satellite systems, as well as spectrally based information acquisition systems in fields other than Earth remote sensing.

A software system for Macintosh and Windows computers called MultiSpec (©Purdue Research Foundation) along with several example data sets are available on a CD as a companion to the material in this textbook. MultiSpec has been used for a number of years with the teaching of the senior/graduate level course mentioned above, and is presently available without cost to anyone via
http://dynamo.ecn.purdue.edu/~biehl/MultiSpec/
MultiSpec contains an implementation of the algorithms discussed in the book. There are 1500 to 2000 downloads of MultiSpec registered

each year, with evidence of many more unregistered downloads. The largest portion of these downloads are to universities for students, classes, and theses in a number of disciplines in countries around the world, but additional copies go into government laboratories, commercial organizations, and the K-12 educational community. It has been licensed to the NASA/NOAA/NSF GLOBE Program
http://www.globe.gov/
which has a membership of 10,000 K-12 schools in 90 countries.

I am greatly indebted to my colleagues, past and present, at Purdue's Laboratory for Applications of Remote Sensing (LARS), which began the study of this technology in 1966. I am especially indebted to Mr. Larry Biehl for his creative work in implementing the algorithms discussed in this book into the MultiSpec applicaton program, and to my former graduate students who taught me a great deal about the characteristics of multispectral and hyperspectral data and how to analyze it.

David A. Landgrebe
West Lafayette, Indiana
October 2002

Part I Introduction

Although this book is printed in black and white, figures where color is necessary or helpful are available for view in color on the accompanying compact disk (CD). The color figures that are available on the CD are noted in the figure captions accordingly.

1 Introduction and Background

This chapter begins with a brief historical background, intended to provide a picture as to how multispectral land remote sensing began and what are its chief motivations. This is followed by a brief system overview and the fundamentals that control how information might be conveyed from the Earth's surface to an aerospace sensor. Next is presented information about the electromagnetic spectrum and the multispectral concept for economically sensing and analyzing measurements. The chapter is concluded with examples showing how data analysis may be carried out using pattern recognition concepts and illustrating one of the less intuitive aspects of the process of information extraction.

1.1 The Beginning of Space Age Remote Sensing

Although remote sensing techniques had been in use for many years, the beginning of the space age greatly stimulated the development of remote sensing technology both for space and for aircraft systems. The space age is generally acknowledged to have begun with the launch of *Sputnik* by the Soviet Union in October 1957. This event had a fundamental effect on society in general. Among other things, it turned thinking to considerations of how satellites might be useful. The launching of *Sputnik* stimulated the U.S. Congress to pass a law in 1958 creating the National Aeronautics and Space Administration (NASA) out of the then-existing National Advisory Committee on Aeronautics (NACA). It took the new NASA only until April 1, 1960, to design, build, and launch TIROS 1, the first satellite designed for Earth observational purposes.

TIROS was a weather satellite designed primarily to provide pictures of the cloud cover over wide areas. However, it was recognized that the ability to operate artificial satellites of the Earth might be useful for many other purposes. In 1966 NASA requested that the National Research Council, the operating arm of the National Academy of Sciences/National Academy of Engineering/Institute of Medicine, undertake a study "on the probable future usefulness of satellites in practical Earth-oriented applications." A steering committee composed of top-level aerospace executives, consultants, and academicians began the work by constructing a list of areas where this use might occur. Then in the summer of 1967 a study group was brought together at the NRC Woods Hole, Massachusetts, study center to examine these possibilities in detail. There were 13 study groups for this work in the following areas:

- Forestry-agriculture-geography
- Geology
- Hydrology
- Meteorology
- Oceanography
- Broadcasting
- Points-to-point communication

- Point-to-pointcommunication
- Navigation and traffic control
- Sensors and data systems
- Geodesy and cartography
- Economic analysis
- Systems for remote sensing
  information and distribution

The 13 panels fall into groups related to remote sensing of the land, atmosphere, and ocean, communication, navigation, and support. The last four–Sensors and Data Systems, Geodesy and Cartography, Economic Analysis, and Systems for Remote-Sensing Information and Distribution–are perhaps best regarded as supporting areas, for they had an impact on all of the other discipline areas. The others were thought to have potential to have a more direct impact on society. The first five involved the development and use of remote sensing technology of the land, the atmosphere and the ocean. The first three were focused on land remote sensing.

In addition to the launching of several meteorology satellites, early research had already begun on land surface sensing by the time of the NRC study. The central question being pursued was could aerospace technology be used to better manage the Earth's natural and man-made resources. The operations related to land remote sensing were expected to have the highest economic impact, because it is on land where the people live that the majority of the economic and resource activity takes place. Thus the potential for the usefulness of land remote sensing technology was thought to be very high.

Of course, the primary driver for the development of such technology is economic. To collect information needed to manage land resources from the ground is slow and time-consuming, and thus expensive. Use of aircraft speeds the process and reduces the cost. Even though an aircraft system is expensive compared to a person walking the ground, aircraft can in fact collect data so much faster and over so large an area that the per unit cost is much lower. Doing so via spacecraft amplifies this cost and speed advantage substantially. However, in each case, aircraft over ground sampling and spacecraft over both, the cost advantage is only realized if there are persons who need the data being gathered. Thus it is important from an economical point of view to have a well-developed user community for such systems, and the systems must be designed to meet what the potential user community perceives as its need. There must also be available viable means for the users to analyze the raw data to provide them with the information they need.

Images (pictures) gathered from spacecraft were early thought of as an important vehicle for collecting information, however, at least among the Sensor and Data Systems panel of the NRC study, the need was discussed for two bases for such systems: imagery-based and spectrally-based systems. Image-based systems would be of the most immediate use if they were obtained at a high enough spatial resolution so that users could "see" details of interest to

5

them in the scene. However, spatial resolution is a very expensive parameter in the space data collection process. To identify a vegetative species as corn, for example, one would require a spatial resolution in the centimeters range so that the shape of a corn leaf could be discerned. Spectrally based methods, which had already begun to be explored by the time of the study, would function best for the same problem with spatial resolutions of the order of a few tens of meters, thus being more economical for data collection and also greatly reducing the volume of data per unit area and thus the processing cost. Spectral methods are the focus of this book. Before proceeding, however, it is interesting to note that, in spite of the NRC anticipated potential of the first three in their list being the highest, all of the projected uses of satellite products are in routine daily use by large segments of the population except those three. The potential in these three areas is still largely ahead of us.

1.2 The Fundamental Basis for Remote Sensing

To begin the discussion of methods for gathering information remotely, consider the fundamental basis for the technology. Information theoretic principles provide that information is potentially available at altitude from the energy fields arising from the Earth's surface, and in particular from the
- Spectral
- Spatial and
- Temporal

variations of those fields. Both the electromagnetic and the gravitational energy fields are of potential interest. To capture the information, one must
- Measure the variations of those fields, and
- Relate them to the information desired.

Though gravity fields are useful for this purpose, we will restrict our considerations to electromagnetic fields.

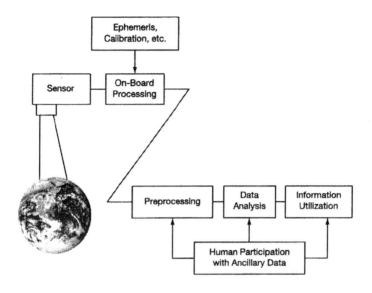

Figure 1-1. A conceptual view of an Earth observational system.

1.3 The Systems View And Its Interdisciplinary Nature.

Figure 1-1 illustrates the organization of an Earth observational system. As the figure shows, the system may be divided into three basic parts: the scene, the sensor, and the processing system.

- The scene is that part of the system that lies in front of the sensor pupil. It consists of the Earth's surface and the intervening atmosphere. For systems in the optical portion of the spectrum that rely on solar illumination, the sun is included. This portion of the system is characterized by not being under human control, not during system design nor operation. However, its most defining characteristic is that it is by far the most complex and dynamic part of the system. It is easy to underestimate this complexity and dynamism.

7

- The sensor is that part of the system that collects the main body of the data to be used. In order to simplify the discussion, we will focus on passive, optical systems, although active microwave (radar) and optical (lidar) systems are important possibilities, in which case the illumination mechanism might be regarded as part of the sensor system. The sensor system is usually characterized by being under human control during the system design phase, but less so or not at all during system operation.

- The processing system is that part of the system over which the data analyst is able to exercise the most control and have the most choices. Although there are some limited circumstances where complete automation is both possible and desirable, this is not the case for most uses. There are several reasons for this. First, it is highly desirable to bring the exceptional human associative and reasoning powers to bear on an analysis problem as a complement to the quantitative powers of computing devices. Since they are complementary, better results should be possible by cooperatively applying both. However, much of the success of an analysis operation hinges on the analyst being able to accurately and thoroughly specify what is desired from the analysis in the face of that exceptional complexity and dynamism of the Earth's surface. One might divide the spectrum of possible degrees of human/machine participation in the analysis process into four parts: fully manual methods, machine aided manual methods, manually aided machine methods, and automatic analysis. Though there are all sorts of variations and specifications that may be present for a given analysis, the center point of this range of uses is perhaps best described as a "manually aided machine analysis."

8

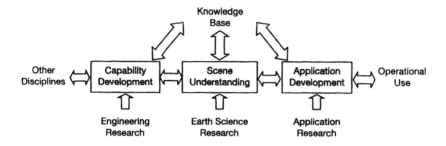

Figure 1-2. The research process for land remote sensing technology.

The research to create such technology is inherently cross-disciplinary. Figure 1-2 illustrates this process. Engineering research is required to create the basic sensor, the data system, and the analysis capability. Earth science research adds the understanding of Earth surface scenes and their spectral characteristics. Researchers familiar with how the needed data are to be utilized are needed to ready the capability for operational use. Science oriented research causes flow upward to increase the knowledge base of the technology while engineering research uses that knowledge base to move understanding and capability to the right toward actual use. This, of course, is the actual goal whether it be use for Earth science research itself or for application. Note that the arrows are all double-ended implying that, especially in the technology creation process, the process should be as integrated as possible.

1.4 The EM Spectrum and How Information Is Conveyed

Earlier it was stated that information is conveyed in the spatial, spectral, and temporal variations of the electromagnetic field arising from a scene. Let us examine this statement more closely. Consider

9

Figure 1-3. Air view of a scene.

Figure 1-3. What information can be derived from this image? First, one sees many straight lines in the image, indicating that there is human activity in the area. One also can see small angular shaped objects in the scene, farmsteads, indicating the humans not only work but live in the area. This is obviously an agricultural area, and these few pieces of information derived from the image of the area were based on the spatial variations apparent in the image.

Now consider Figure 1-4, which gives the same scene of Figure 1-3 in color. One now can derive additional pieces of information about the scene. For example, some of the areas are green, indicating an emerging canopy of a crop, while others areas are brown, indicating a crop ready for harvest or having been harvested. If one had in hand certain ancillary data such as the longitude and latitude of the scene (this scene is from the mid-latitude U.S. Midwest), along with some agricultural knowledge, one might even be able to draw the conclusion that the green fields are corn or soybeans and the brown fields are wheat or other small grain, and that the

Figure 1-4. A color air view of the scene of Figure 1-3. (In color on CD)

scene was imaged in late June or early July. The availability of the color in the image adds a small measure of the spectral variations arising from the scene, and this allows for additional information to be derived over the spatial variations only of Figure 1-3.

Next consider Figure 1-5. This is a color image of the same scene, only using a different type of color film known as color infrared or color IR film. Color IR film came to prominence during World War II when it was called camouflage detection film. Figure 1-6 shows a conceptual drawing of the spectral reflectance of green vegetation, and how color and color IR film relate to it.

Figure 1-5. The scene of Figures 1-3 and 1-4 but in color IR form.
(In color on CD)

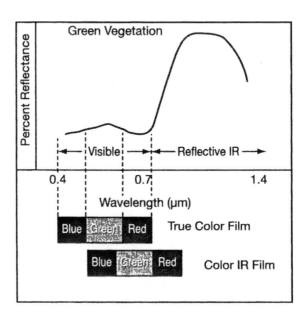

Figure 1-6. The concept of color IR film. (In color on CD)

Note that the relative reflectance for vegetation is higher in the visible green region than in the blue or red, thus giving the vegetation the visible green color. However, note that the reflectance is much higher in the reflective infrared region than anywhere in the visible. This high IR reflectance is something unique to green vegetation as compared to, for example, green netting thrown over a tank, thus its camouflage detection property. Color IR film was devised to take advantage of this by shifting the sensitivity of the dye layers of the color film one notch to the right as shown in the figure.

Since the red dye layer of the film is sensitized to the high reflectance in the infrared, true green vegetation would show as red in color IR film. Because of the high infrared response of green vegetation, in nonmilitary circumstances it has proved useful in displaying the degree of health, vigor, and canopy cover of green vegetation, and has become a standard in the remote sensing field. It is used here to demonstrate a case where looking beyond the visible portion of the spectrum can significantly augment the information-gathering potential of spectral variations. The point of these images is to show intuitively that spatial and spectral variations do indeed bear information that may be useful. Temporal variations will be discussed later.

Figure 1-7 shows the layout and principle nomenclature for the entire electromagnetic spectrum. Of particular interest in remote sensing are the optical and microwave wavelengths. In this diagram, the optical portion is expanded to show the major sub-areas. Note that the optical regions useful for remote sensing extend beyond that where photography can be used; thus other types of sensor systems must be used. More details on the optical spectrum will be given in the next chapter.

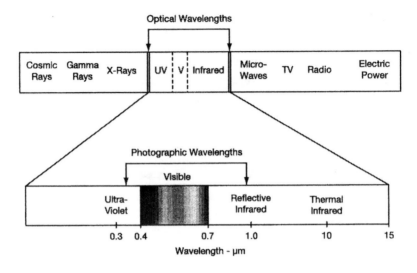

Figure 1-7. The Electromagnetic Spectrum. (In color on CD)

1.5 The Multispectral Concept and Data Representations

The energy field arising from the Earth is, of course, finite in magnitude. The data collection process must divide this finite quantity spatially into pixels. The power level in each pixel can be divided up into a number of spectral bands. Given the finite nature of the field, there is then a tradeoff between these two, because as one moves to finer spatial resolution and spectral band intervals, less power is left to overcome the internal noise inevidently present in the sensor system. Thus a less precise measure can be made of the signal level arriving from the surface. This demonstrates that the fundamental parameters of this information-gathering process are

- The spatial resolution,
- The spectral resolution, and
- The signal-to-noise ratio (S/N)

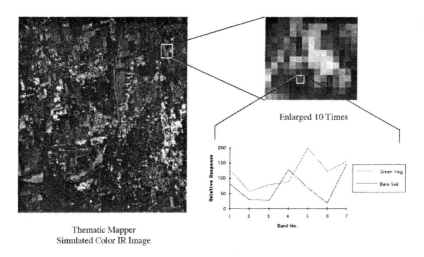

Enlarged 10 Times

Thematic Mapper
Simulated Color IR Image

Figure 1-8. The multispectral concept, showing the signal from a pixel expressed as a graph of response vs. spectral band number. (In color on CD)

and that they are all interrelated. Now the power level decreases and the volume of data that must be delt with increases both as the square of the spatial resolution, but only linearly with spectral resolution. A major motivation for using spectral variations as the primary source from which to derive information is therefore to avoid the need for very high spatial resolution.

Figure 1-8 illustrates the multispectral concept for information extraction. On the left is shown a simulated color IR image created from Thematic Mapper data for a small urban area. Thematic Mapper is a Landsat spacecraft instrument that collects data in seven spectral bands from pixels that are 30 meters in size and with a S/N providing 8-bit ($2^8 = 256$ levels of measurement precision) data. The image is a systematic presentation of data from three of the seven bands. The small region marked is shown expanded 10 times

15

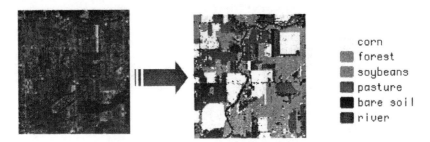

corn
forest
soybeans
pasture
bare soil
river

Figure 1-9. A thematic map of an agricultural area created from Thematic Mapper multispectral data. (In color on CD)

in the upper right, making the individual pixels visible. Below this is a graphical presentation of the seven measurements made on the individual pixel marked, compared to a different pixel containing bare soil. Using the multispectral concept, it is from these seven numbers that the assignment of those pixels would be made to one of an established list of classes.

Figure 1-9 provides an example result that this process could produce. In this so-called thematic map presentation, each pixel of the data for the region shown in the color IR image has been assigned to one of the six classes on the right using the color code shown. One can see subjectively what the distribution of land cover classes looks like from the thematic map and one could obtain a quantitative measure of the amount of each land cover class by counting the number of pixels assigned to each class.

Summarizing to this point, it was hypothesized that information should be available at the sensor pupil via the spectral, spatial, and temporal variations that are present in the electromagnetic fields emanating from the Earth's surface. For the present, we are considering measurements made at a single instant of time, and will postpone the consideration of temporal variations until later. We have examined conceptually (but rather superficially so far) the hypothesis of information from spatial and spectral variations, and found at least some rationale for the validity of both. Let us consider both in a bit more detail.

As suggested by the preceding example, if the pattern of variations across several spectral bands for a given Earth cover type can be shown to be diagnostic, this can serve as a basis for discriminating between cover types. The required spatial resolution can be substantially reduced to the point of pixel sizes large enough to measure the net spectral response, for example, of several rows of corn instead of needing to be fine enough to see the shape of the corn leaves. Though not as intuitive as spatial methods, this turns out to be quite practical. Thus a more fundamentally rigorous form of this line of thinking is needed.

To begin such a more fundamentally rigorous formulation, consider Figure 1-10. On the left is the radiance response as a function of wavelength for three example classes of surface cover. Suppose one has an instrument that samples the responses at two different wavelengths as indicated. Then the data could be plotted in a two-dimensional space as indicated on the right, the magnitude of the response in one band vs. that in the other. Since the three materials

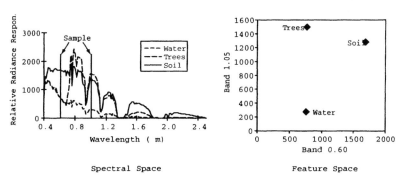

Figure 1-10. Data in spectral space and in feature space. (In color on CD)

17

have different responses at the two wavelengths, the points in this new space, called a feature space, plot in different parts of the space.

Such a feature space proves to be useful for establishing a computer algorithm for discriminating among spectral responses. Note that the samples for trees and water were not separable with only the band at 0.60 μm, but by adding the second band, they became separable. If one had a case were materials were not separable with two bands, one could add a third or a fourth, or more and eventually achieve separability. Notice also that one of the unique things about the feature space representation is that by adding more bands, the responses remain points in feature space, only the dimensionality of the space increases. The original data in spectral space was actually a plot of the three materials at 210 different wavelengths from 0.4 to 2.4 μm. Thus, if one went to a 210-dimensional feature space, one could represent all of the information the spectral space contained, and the three graphs of spectral space would have been converted to three points in 210-dimensional space. Without losing any information in the process, a graph in spectral space is converted to a point in feature space.

Thus we have introduced three different types of signal representations or signal spaces. Data represented in image space proves very useful to locate geographically where a pixel is from, but its effectiveness does not extend beyond two or at most three dimensions (spectral bands). One cannot in effect display signal variations for more than three bands at once. Data can also be expressed in spectral space. This type of representation is useful in relating a given pixel's response to the physical basis for that response and for any number of spectral bands. But it is difficult to display all of the diagnostic variation in a spectral space form. The third is feature space where it is possible to represent all of the diagnostic information mathematically, and therefore for

18

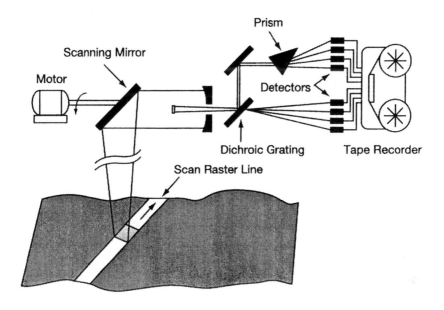

Figure 1-11. Diagram of a scanner that collects data
directly into feature space.

computation, but one cannot show this graphically for human view.
We will deal more fully with these three in chapter 3.

Before seeking means for analyzing data represented in feature
space, consider briefly how data may be collected for such a feature
space. See Figure 1-11. This is a diagram of one type of device
(there are many) for collecting multispectral image data that is
directly in feature space. The scanning mirror sweeps across the
field of view as the platform (aircraft or spacecraft) flies along to
the left. If one stops the scanning for an instant to follow the energy
being collected, one sees that the radiation from a single pixel is
directed through some optics to a dispersive device, in this case a
prism or a grating, which spreads out the energy from the pixel
according to its wavelength. The radiation from each wavelength
is fed to its own detector for conversion to an electrical signal that
is recorded or directly transmitted to the ground. Thus, data from

19

all bands for a pixel are simultaneously collected to form a vector in feature space. Then, the scanning mirror sweeps across the field of view picking up the radiation from a row of pixels sequentially, and as the platform moves forward, the scanning mirror picks up radiation for the next row of pixels. In this way the entire scene can be rastered, with a multispectral vector generated for each pixel.

1.6 Data Analysis and Partitioning Feature Space

The basic idea of pattern recognition methods for feature space data is to partition up the entire feature space into M exhaustive, nonoverlapping regions, where M is the number of classes present in the scene, so that every point in the feature space is uniquely associated with one of the M classes. Then any measurement result may be automatically assigned to one and only one of the classes.

There are many algorithms for making this partitioning. A very simple one can be constructed as follows. Consider Figure 1-12. Suppose that one wishes to classify a data set into three classes. Suppose also that one can somehow obtain a small number of representative samples for each class for which the correct labels are known. These samples are referred to as training samples or design samples, as it is from these samples that the partitioning of the feature space will be

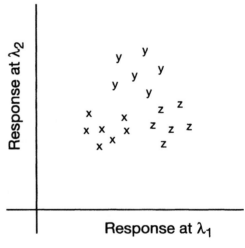

Figure 1-12. Hypothetical data in 2-space

determined. Assume that the samples shown in Figure 1-12 are the samples to be used for training the classifier.

Using the training samples, one can calculate the class mean for each of the three classes, and then draw the locus of points that are equidistant from each pair of mean values, as shown in

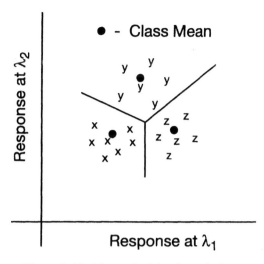

Figure 1-13. Linear decision boundaries.

Figure 1-13. This in effect defines decision boundaries for each of the class pairs so that the desired partitioning of the feature space is

defined. Then a computer algorithm can be used to assign any unknown measurement to one of the three classes. This particular classifier is known as a "minimum distance to means" classifier.

As previously noted, there are many ways to achieve the partitioning of the feature space, some

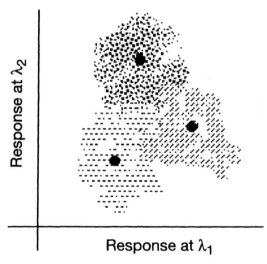

Figure 1-14. A more typical situation for class distributions.

21

resulting in linear or piecewise linear decision boundaries and some resulting in nonlinear boundaries. There are a number of additional factors involved in deciding which is the best algorithm to use in any given case. However, we note here that the conceptual drawings of Figures 1-12 and 1-13 are much simpler than the usual case in practical problems. Usually data are not so obviously separable as was the case in these examples. A more typical, though still hypothetical, situation might be as shown in Figure 1-14. As is seen here, the optimal location or shape for the decision boundaries is not obvious. Furthermore one must nearly always use at least four and usually more features to achieve satisfactory performance. Thus the analyst may not be able to "see" the data and how it distributes itself in a higher dimensional space because one cannot see in high-dimensional spaces. It is a matter of being able to make calculations to determine the optimal boundary locations, rather than being able to "see" where they should be. In the next section some additional basic characteristics will be presented in a conceptual rather than mathematical form.

1.7 The Significance of Second-Order Variations

In the preceding we have shown a very simple classifier that represents each class by a single point,namely the class mean value, which is in effect, a single spectral curve. However, this is a rather incomplete description of a class. How the spectral response varies about its mean value also is quite information-bearing. To examine this aspect, we willreturn to image space to make clear what looking at only images and the average brightness can overlook. Consider Figure 1-15. These image space presentations are of the spectral response in 12 spectral bands for an agricultural area. The question to be asked is, Which bands are useful for discriminating among the various crops being imaged? There does not seem to be a way to answer this question, however, as there is too much information

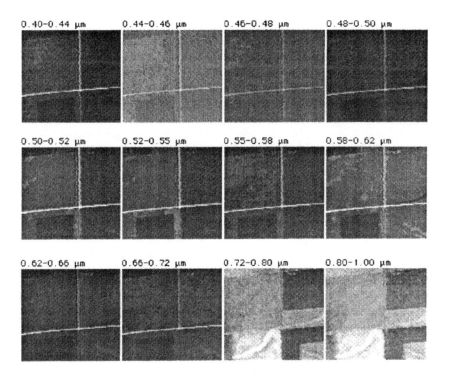

Figure 1-15. Images in 12 spectral bands.

to be perceived from an image space presentation of the data. Rather, the problem does introduce an important aspect of classification, namely how does one choose the spectral bands to be used for classification. As will become apparent later, using too many bands, too high a dimensional space, can be just as bad as not using enough.

Therefore, for now, let us simplify the problem. See Figure 1-16. Suppose that the question is reduced to, Having decided to use the 0.52-0.55 and 0.55-0.58 μm bands, would there be any value to adding the 0.50-0.52 μm band? From the images, it would appear that the data in the 0.50-0.52 μm band is highly correlated with that in the 0.52-0.55 μm band. A common view of correlation is

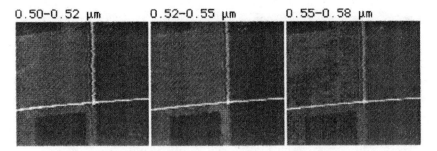

0.50-0.52 μm   0.52-0.55 μm   0.55-0.58 μm

Figure 1-16. Images fom three spectral bands.

that correlation implies redundancy, and so given one, the other doesn't imply additional information. However, this turns out to be an inappropriate viewpoint.

Instead of exploring this kind of a problem in image space, let us turn to spectral space. Consider Figure 1-17. This is the presentation of the composite of a number of laboratory spectral measurements of the percent reflectance of the leaves of two classes of vegetation. Do the two classes represented there appear to be separable? From

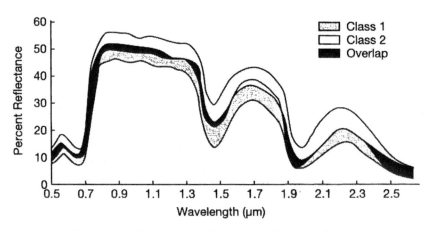

Figure 1-17. Two classes of vegetation in spectral space.

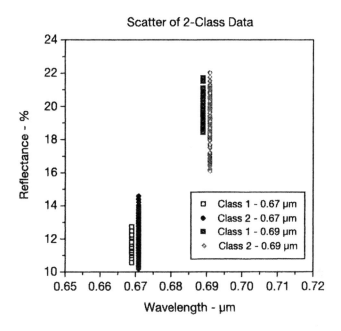

Figure 1-18. Scatter plot of the two classes of Figure 1-17
in the 0.67 and 0.69 μm bands.

looking at the curves, the answer would appear clearly to be yes. Simply set a level at about 38% reflectance in the 1.7 μm band.

However, let us make the problem more difficult. Suppose that the data collection circumstances constrained one to use no more than two spectral bands, but in the vicinity of 0.7 μm. Now, are the classes separable? Viewing the data in spectral space as in Figure 1-18, it would certainly appear that the data for the classes is overlapping in both bands, and so they are not separable.

Let us now consider the same data, but in feature space, as shown in Figure 1-19. Viewed in this way it is apparent that the two classes are quite separable, even with a linear classification algorithm. Note that the data in both bands are overlapping when

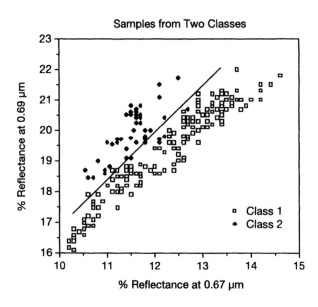

Figure 1-19. The two classes of vegetation and two bands in feature space.

the shape of the two class distributions in two dimensions shows a high degree of positive correlation between the two bands. This is indicated by the tendancy of both classes to lie along lines at 45° upward to the right passing through the class mean. It is also the case that the higher the correlation, the more closely the data will fall near to the 45° lines. Thus this high correlation between bands coupled with only a slight difference in mean value locations makes the two classes separable. It may be concluded from this that rather than indicating redundancy, the interest in correlation in this context is that it provides information about the shape of the class distriabution. Thus the mean value, a first-order statistic, provides location information about a class in feature space, while a second-order statistic, correlation, provides information about the shape of the class distribution. Either or both can provide characteristics useful in separating classes.

1.8 Summary

In this chapter the basic concepts of multispectral land remote sensing were introduced. After presenting a bit of history and the basic fundamentals of the field from an information theoretic standpoint, the multispectral approach was described in conceptual form as an alternative to image processing methods. The intent is to create a capability that is effective but economical to utilize, since one of the great advantages of remote sensing is that it enables the gathering of information useful in managing the Earth's resourses in a way that takes advantage of the economies of scale. Although the multipectral concept is perhaps less intuitive than reliance on images, it is simpler and more directly suited to computer implementation. The basic concepts of feature space representations and the use of pattern recognition methods to provide the needed discrimination among classes were outlined. And finally, a hint of the power of these methods, which perhaps exceed one's intuition, was given with the simple two-class, two-dimensional example.

Thus several concepts of multispectral technology were outlined in this first chapter. What remains is a more thorough detailing of these concepts to achieve a really practical, usable technology.

Part II The Basics for Conventional Multispectral Data

Although this book is printed in black and white, figures where color is necessary or helpful are available for view in color on the accompanying compact disk (CD). The color figures that are available on the CD are noted in the figure captions accordingly.

2 Radiation and Sensor Systems in Remote Sensing

In this chapter, basic information is presented about the illumination of Earth surfaces and sensor systems to measure the power emanating from them. After an introductory overview of the basic system parts, radiation terminology and spectral nomenclature are introduced, as well as Planck's radiation law and the fundamentals of black body radiation. This is followed by a discussion of atmospheric effects on the inbound and outbound radiation, including the adjacency effect that arises due to atmospheric scattering. Next is introduced the basics of optics and then means for quantitatively describing surface reflectance. An overview of radiation detectors follows. The chapter is concluded by integrating these elements into complete multispectral sensor systems, factors that provide distorting influences in the collection of data, and descriptions of some example sensor systems that have been historically significant.

2.0 Introduction[1]

We begin with a brief summary of system variables related to data acquisition, including the major fundamentals of radiation, the sun as an illumination source, and sensor system technology. The discussion will be focused on passive optical systems where the source of illumination is the sun, although many of the principles

---

1   Some of the material of this chapter is a revised and extended version of the contents of Chapter 2 of P.H. Swain and S.M. Davis, editors, *Remote Sensing: The Quantitative Approach*, McGraw-Hill, 1978. That chapter was originally written by Prof. LeRoy Silva of Purdue University.

discussed would also apply to active systems such as those using lidar or those in the microwave region such as radar.

There are four basic parts to a data acquisition system,
- the radiation source,
- the atmospheric path,
- the Earth surface subject matter, and
- the sensor

See Figure 2-1. Earlier it was noted that potentially useful information is available from the electromagnetic energy field arising from the Earth's surface via the spectral, spatial, and temporal variations in that field. The interaction of the illumination energy with the target is where the information signal originates. Thus factors that affect this energy field are significant to the information acquisition process, and it is useful to review each of the four parts.

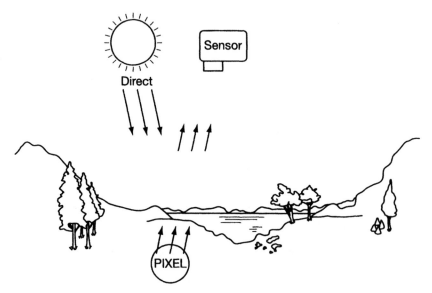

Figure 2-1. A conceptual view of the scene.

2.1 Radiation Terminology and Units

Any body whose temperature is above absolute zero radiates electromagnetic energy. Bodies may also transmit or reflect a portion of the radiation that falls upon them from other sources. The magnitude of this radiation varies with wavelength and depends on the illumination of the body, the temperature of the body, and upon some of its physical characteristics as well. At the outset, it is useful to have an appropriate system of terms and units with which to refer to the radiation in the various circumstances of remote sensing. Photometric terminology and units, which relate to the human eye, are one possible basis for such a system of units. But for the work here, the human eye is not the sensor system in mind. Radiometric

Symbol	Description	Defining Expression	Units	Units (Abbreviation)
Q	Radiant energy		Joules	J
Φ	Radiant flux (power)	$\dfrac{dQ}{dt}$	Watts	W
E	Irradiance	$\dfrac{d\Phi}{dA}(in)$	Watts per square meter	W/m²
M	Radiant Exitance	$\dfrac{d\Phi}{dA}(out)$	Watts per square meter	W/m²
L	Radiance	$\dfrac{d^2\Phi}{\cos\theta\, dA\, d\omega}(out)$	Watts per square meter per steradian	$\dfrac{W}{(m^2-sr)}$
E_λ	Spectral Irradiance	$\dfrac{dE}{d\lambda}(in)$	Watts per square meter per micrometer	$\dfrac{W}{(m^2-\mu m)}$
M_λ	Spectral Radiant Exitance	$\dfrac{dM}{d\lambda}(out)$	Watts per square meter per micrometer	$\dfrac{W}{(m^2-\mu m)}$
L_λ	Spectral Radiance	$\dfrac{dL}{d\lambda}$	Watts per square meter per steradian per micrometer	$\dfrac{W}{(m^2-sr-\mu m)}$

Table 2-1. A radiometric system of units for remote sensing quantities.

terminology and units are more absolute in character, thus are more appropriate to the use here. Table 2-1 provides such a system.

In Table 2-1, radiant exitance implies radiance leaving a surface. Irradiance implies incident radiance. Note that the quantity L (Radiance) has geometry associated with it. The reason for this is as follows: If an object that does not fill the field of view is viewed at normal incidence, it has a certain apparent area. This apparent area decreases as one moves away from normal in an angular fashion, as shown in Figure 2-2, thus the cosine in the denominator of the definition.

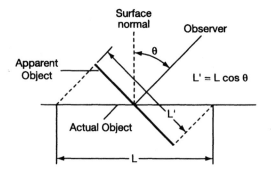

Figure 2-2. Projected area effects for an object that does not fill the field of view.

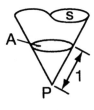

Figure 2-3. Solid angle defined.

Solid angle measure occurs in some of the terms of Table 2-1. As a brief review of solid angle, which is measured in steradians, the solid angle Ψ at any point P subtended by a surface S is equal to the area A of the portion of the surface of a sphere of unit radius, centered at P, which is cut out by a conical surface, with vertex at P, passing through the perimeter of S. See Figure 2-3. Recall also that the area of the surface of a sphere is $4\pi r^2$. Since the sphere of which A is a part has unit radius, there is a total of 4π steradians in a sphere.

The last three terms of Table 2-1 are functions of λ (wavelength), that is, they describe how energy is distributed over the spectrum. Notice how the independent variables of Table 2-1, λ, A, and t, correspond to the spectral, spatial, and temporal variables of the electromagnetic field, the relevant variables for information available from radiation fields.

2.2 Planck's Law and Black Body Radiation

When radiation falls on a body, the radiation may be reflected, transmitted, or absorbed by the body, and if absorbed, the energy is converted to heat and can be re-radiated by the body at another wavelength. However, the Law of Conservation of Energy applies. Thus,

$$\rho + \alpha + \tau = 1$$

where ρ, α, and τ are the ratios, respectively, of the reflected, absorbed, and transmitted energy to the incident energy. Thus all of the radiation must be accounted for in those three mechanisms, reflection, transmission, and absorption.

A black body is an ideal radiator, implying that it transmits and reflects nothing and thus is an ideal absorber. Black body radiation is governed by Planck's Radiation Law:

$$M_\lambda = \frac{\varepsilon\, c_1}{\lambda^5 \left(e^{c_2/\lambda T} - 1 \right)} \quad W/(m^2 - \mu m)$$

where M_λ = spectral radiant exitance,
           ε = emittance (emissivity), dimensionless
           c_1 = first radiation constant, 3.7413×10^8 W-$(\mu m)^4/m^2$

35

λ = radiation wavelength, μm
c_2 = second radiation constant, 1.4388×10^4 μm-K
T = absolute radiant temperature, $^\circ$K

Values of M_λ for several values of temperature are shown in Figure 2-4. Notice how the peak of these radiation curves move upward (rapidly!) and to the left as temperature increases.

2.3 Solar Radiation

Typical radiation curves for sunlight are shown in Figure 2-5. Note that the solar extraterrestrial and the 6000°K black body curves correspond rather well. The difference between the extraterrestrial

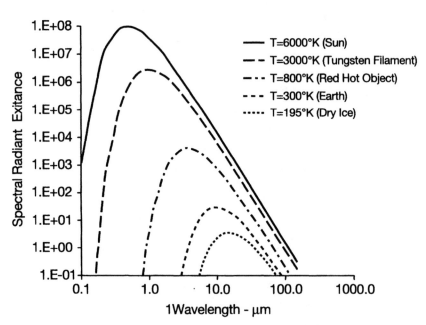

Figure 2-4. Planck's Law radiation for several representative temperatures.

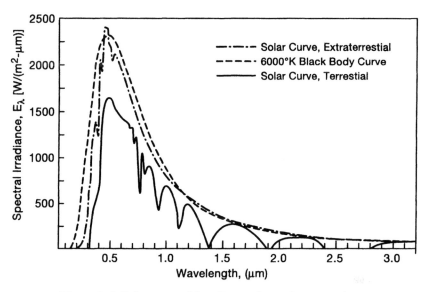

Figure 2-5. Solar spectral irradiance above the atmosphere
and at the Earth's surface through one air mass.

and the terrestrial solar curves implies an atmospheric effect. This
effect is not constant with time or place. It varies from instant to
instant and from square meter to square meter; thus these curves
shown are only representative. It is also seen that the solar
illumination is high in the visible (0.4–0.7 µm), thus making possible
a high signal level in this area. It becomes quite low beyond 3 µm,
thus decreasing or eliminating the utility of this area for passive
remote sensing purposes.

If one integrates Planck's Law over all wavelengths to derive the
equation for M, the radiant exitance, the result is

$$M = \varepsilon\sigma T^4 \qquad W/m^2$$

37

where $\sigma = 5.6693 \times 10^{-8}$ W/(m²-K⁴). This is known as the Stefan-Boltzmann radiation law. Further, if Planck's law is differentiated and the result set to 0 to find the λ_{max} the result is

$$\lambda_{max} = \frac{2898}{T} \mu m \text{ (T in °K)}$$

Thus, for example, the sun as a 6000°K blackbody has its maximum at 0.483 μm, which is in the visible portion of the optical region.

Note that an object at room temperature ($\approx$ 300°K) would have spectral radiant exitance as in Figure 2-6, this time given on a linear scale. The exitance is seen to peak at 9.66 μm and most of its power is between 7 and 15 μm. That is what makes this region, the so-called thermal region, potentially useful in Earth remote sensing.

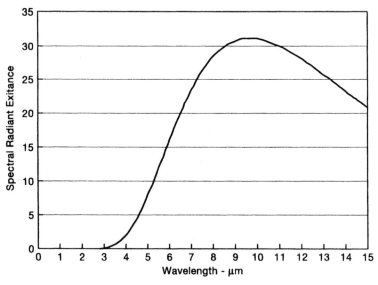

Figure 2-6. Black body emission for a 300°K body.

Because materials are often opaque in the thermal region, one can frequently interpret radiation in the 7-15 μm region in terms of temperature. A common use of this idea is in remotely sensing water temperature, since water is known to be opaque in the thermal region. However, note that one is sensing the surface skin temperature, not the bulk temperature, and these temperatures may be quite different.

Planck's Law assumes the radiator is an ideal black body. The deviation of the constant ε from unity is a measure of deviation from the ideal. It is called the emittance, or more commonly the emissivity. Targets may have emissivities that are not constant with λ. If a target does have emissivity that is constant with λ but not equal to unity, the body is referred to as a gray body. Thus we have

Black body $\varepsilon = 1$
Gray body $0 < \varepsilon < 1$
Perfect reflector $\varepsilon = 0$
All others $\varepsilon = f(\lambda)$

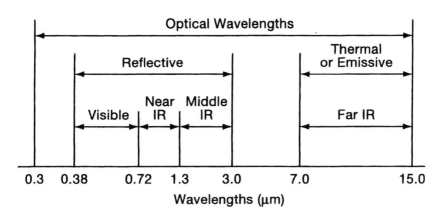

Figure 2-7. Standard nominclature for the various parts of the optical spectrum.

Alternately, one may account for differences between black body exitance and that of a given natural target by assuming $\varepsilon = 1$, but its temperature is a function of wavelength.

Figure 2-7 provides some standard nomenclature frequently used for the various parts of the spectrum. Notice that the region from 3.0 to 7.0 μm is not given a special name in this figure. In this region the illumination from the sun has fallen off to a nearly negligible level so that the reflective mechanisms are no longer useful, and the thermal radiation of bodies at or near room temperature has not yet risen to a useful level.

2.4 Atmospheric Effects

Scattering

The atmosphere affects optical radiation in two ways, scattering and absorption. However, the processes are not simple, and not constant over space or time. Both are wavelength dependent. See Figure 2-8. Scattering occurs when radiation is reflected or refracted by atmospheric particles. Example scattering particles are gas molecules, dust particles, and water molecules. Scattering is a re-direction of energy, and is usually divided into three categories, depending on the size of the scattering particle, as follows.

1. **Rayleigh scattering** occurs when the radiation wavelength is much larger than the size of the scattering particle. Some of the factors influencing the scattering process can be seen from the equation for the Volume Scattering Coefficient, indicating the relative amount of the radiation that is scattered, as follows.

$$\sigma_\lambda = 4\pi^2 \frac{NV^2(n^2 - n_0^2)^2}{\lambda^4(n^2 + n_0^2)^2}$$

where

N - number of particles/cm³
V - volume of scattering particles
λ - radiation wavelength
n - refractive index of particles
n_0 - refractive index of medium

Because of the inverse λ^4 term, shorter wavelengths are scattered more strongly than longer ones. As a result a portion of the incoming solar rays, especially at the blue end of the spectrum, are scattered over the whole sky. This is what causes the sky to appear blue. It is also the cause of red sunsets in that the low sun angle results in a long path for solar rays. As a result most of the short wavelengths are scattered, leaving only the red to reach our eyes. As a result of Rayleigh scattering, the blue end of the visible spectrum is less useful to remote sensing, and the ultra violet region is seldom used at all.

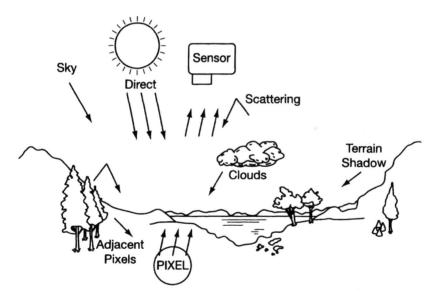

Figure 2-8. Some of the observation variables that affect the measured spectral response.

2. The second type of atmospheric scattering is called **Mie scattering**. It occurs when the wavelength is comparable to the size of the particles. In this case the scattering coefficient is given by

$$\sigma_\lambda = 10^5 \pi \int_{a_1}^{a_2} N(a)K(a,n)a^2 da$$

N(a) - number of particles in (a, a+da)
K(a,n) - scattering cross section
a - radius of spherical particles
n - index of refraction of particles
Mie scattering is most commonly a problem due to haze.

3. The third type is referred to as **Nonselective scattering**. It occurs when scattering particles are much larger than the wavelength. It is the net effect from reflection by the particle, passage though the particle, and refraction at the particle edge. It becomes a factor when the atmosphere is dust-laden

Absorption is the conversion of energy at a given wavelength to heat. It occurs also at the molecular level, with molecular resonance, thus resulting in very narrow "absorption bands." In this case, it becomes dominant over other mechanisms.

In sum, three things can happen to an optical ray in the atmosphere—scattering, absorption, and transmission. Notice also that the atmosphere is not uniform, but its density decreases with altitude. Thus the effect on inbound rays is not the same as that on outbound rays. It is also certainly not uniform horizontally.

Transmittance and Path Radiance

The identifiability of a target is a property of its reflectance. What can be measured is the spectral radiance at the sensor pupil. The relation between target reflectance and pupil spectral radiance depends on target irradiance and on the atmosphere. In fact these two effects are dependent upon a number of other things.

What is the relationship between the radiant exitance from the surface with that that emerges from the top of the atmosphere? A common first-order approximate model used to quantify this effect is

$$L_\lambda = \tau_a(\lambda)\, L_s(\lambda) + L_p(\lambda)$$

where

$L_\lambda(\lambda)$ - spectral radiance emerging from the top of the atmosphere

$L_s(\lambda)$ - spectral radiance entering the atmosphere upward from the surface

$\tau_a(\lambda)$ - atmospheric transmittance, a ratio expressing the portion of the spectral radiance that is transmitted through the atmosphere

$L_p(\lambda)$ - path radiance.

Path radiance (up-welling radiation) is radiation entering the sensor aperture that did not arrive directly from the scene pixel under view. Its origin is scattering from the atmosphere and radiance from adjacent scene pixels.

Transmittance is related to Visibility or Meteorological Range, a quantity often available in a generalized sense from meteorological sources for a given area. It is related to transmittance in the following fashion: Transmittance from the earth's surface into the outer atmosphere is given by

$$\tau_a(\lambda, V_\eta, \theta) = \varepsilon^{-\tau'_{ext}(\lambda, V_\eta)\sec\theta}$$

where

τ_a - atmospheric transmittance

τ'_{ext} - extinction optical thickness (from Rayleigh and aerosol scattering and ozone absorption)

V_η - visibility or meteorological range

θ - solar zenith angle

Table 2-2 relates Extinction Optical Thickness to Meteorological Range.

	Meteorological Range V_η (km)				
λ	2	4	6	8	10
0.27	76.276	74.825	74.299	74.020	73.847
0.28	40.658	39.289	38.794	38.530	38.368
0.30	7.657	6.364	5.896	5.647	5.493
0.32	4.080	2.863	2.422	2.187	2.042
0.34	3.414	2.273	1.859	1.639	1.503
0.36	3.086	2.014	1.625	1.418	1.291
0.38	2.881	1.847	1.472	1.272	1.149
0.40	2.593	1.642	1.297	1.114	1.000
0.45	2.203	1.360	1.054	0.891	0.791
0.50	1.968	1.197	0.917	0.768	0.676
0.55	1.805	1.092	0.834	0.696	0.611
0.60	1.624	0.984	0.751	0.627	0.551
0.65	1.452	0.868	0.655	0.542	0.473
0.70	1.341	0.793	0.594	0.488	0.423
0.80	1.178	0.692	0.516	0.422	0.364
0.90	1.067	0.624	0.463	0.378	0.325
1.06	0.961	0.561	0.416	0.338	0.290
1.26	0.876	0.511	0.379	0.308	0.264
1.67	0.753	0.441	0.326	0.266	0.228
2.17	0.672	0.392	0.290	0.236	0.202

Table 2-2. Extinction Optical Thickness $\tau'_{ext}(\lambda, V_\eta)$ as a function of wavelength and meteorological range.

Using Table 2-2 and the preceding relationship, one obtains an example table of transmittance for various meteorological ranges and wavelengths for a solar zenith angle of 20° shown in Table 2-3. (See also the atmospheric transmittance graphs of Figure 2-9.)

	Meteorological Range V_η (km)				
λ	2	4	6	8	10
0.27	0.00	0.00	0.00	0.00	0.00
0.28	0.00	0.00	0.00	0.00	0.00
0.30	0.00	0.00	0.00	0.00	0.00
0.32	0.01	0.05	0.08	0.10	0.11
0.34	0.03	0.09	0.14	0.17	0.20
0.36	0.04	0.12	0.18	0.22	0.25
0.38	0.05	0.14	0.21	0.26	0.29
0.40	0.06	0.17	0.25	0.31	0.35
0.45	0.10	0.24	0.33	0.39	0.43
0.50	0.12	0.28	0.38	0.44	0.49
0.55	0.15	0.31	0.41	0.48	0.52
0.60	0.18	0.35	0.45	0.51	0.56
0.65	0.21	0.40	0.50	0.56	0.60
0.70	0.24	0.43	0.53	0.59	0.64
0.80	0.29	0.48	0.58	0.64	0.68
0.90	0.32	0.51	0.61	0.67	0.71
1.06	0.36	0.55	0.64	0.70	0.73
1.26	0.39	0.58	0.67	0.72	0.76
1.67	0.45	0.63	0.71	0.75	0.78
2.17	0.49	0.66	0.73	0.78	0.81

Table 2-3. Atmospheric Transmittance τ_a $(\lambda, V_\eta, \theta)$ for a Solar Zenith Angle of 20°.

From Table 2-3, one can clearly see how the atmospheric transmittance falls off at the shorter wavelengths (mostly due to scattering) and for less clear atmospheres (shorter meteorological ranges). This decrease can substantially reduce the utility at those wavelengths and in such weather conditions by significantly reducing the signal-to-noise ratio out of the sensor.

The total target illumination depends on direct solar rays, atmospheric transmission in the downward path, and secondary illumination sources, such as sky radiance, and reflected radiance from nearby bodies such as clouds and surface bodies.

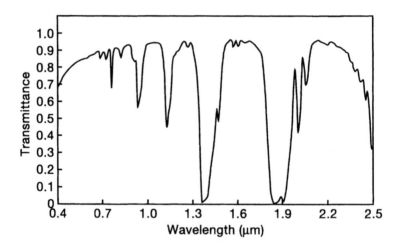

Figure 2-9. Example of atmospheric transmittance vs. wavelength.

Sky Radiance

In addition to direct solar rays, the surface target is illuminated by the diffuse light from the rest of the sky. This light emanates from direct rays scattered by the atmosphere. This illumination is not spectrally similar to the direct rays. Rather, it tends to be blue on a clear day since it's origin is primarily Rayleigh scattering, but it varies significantly depending on the clarity or optical depth of the atmosphere. As a result the total irradiance falling on a given area can vary quite significantly both in magnitude and spectral distribution depending on the atmosphere. It also varies depending on how much of the sky is visible from the ground surface element under consideration. In a flat area, for example, this can approach a full hemisphere of 2π steradians, while in a valley between mountains it would be substantially less. Table 2-4 gives the proportion of downward diffuse solar flux to total direct downward flux for three different atmospheric conditions at $\lambda = 0.65$ µm, $\alpha = 10\%$, and a zenith angle of 29.3°. Note that the effect also varies significantly with altitude.

Altitude	Atmospheric Condition Clear $\tau = 0.13$	$V_0 = 23$ km $\tau = 0.35$	$V_0 = 5$ km, $\tau = 1.33$
Top of Atmos.	0	0	0
30 km	0.04	0.06	0.07
9 km	1.0	1.9	2.0
2 km	2.6	11.6	12.1
1 km	2.9	17.6	34.1
Surface	3.3	26.0	64.4

Table 2-4. Proportion of downward flux to total direct downward flux for 3 atmospheric conditions and several altitudes

Adjacency Effect

A significant problem arises due to the scattering of the response from nearby objects that have different reflectance characteristics than the object being viewed. A specific example will make this clearer.[2] Assume that one wants to measure the spectral response at an arbitrary point in a body of water that is 2 km across and is completely surrounded by green vegetation. This may be calculated by suitably modeling the scene. Figure 2-10 shows the assumed atmospheric opacity of the multilayered atmospheric model and

---

[2]   This example is due to Dr. David Diner of the NASA/Jet Propulsion Lab. Used by permission courtesy of NASA/JPL/ Caltech

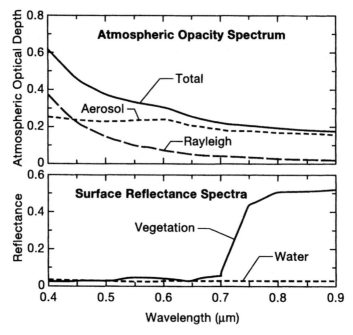

Figure 2-10. Atmospheric opacity and surface reflectance assumed for vegetation surrounding a lake.

the surface reflectance for both the water and the vegetation. Clearly, vegetation and water have very dissimilar spectral characteristics.

Then, if the spectral radiance arriving at a point above the atmosphere is calculated from locations on the water at various distances from shore, the results are as shown in Figure 2-11. Note the large but constant effect at the blue end of the spectrum due principally to Rayleigh scattering. Because this effect is quite constant over space, its principal impact is just to lower the signal-to-noise ratio, and it is otherwise not a serious problem.

However, a more serious problem is the significant extent to which the vegetation reflectance gets mixed into the response for the water

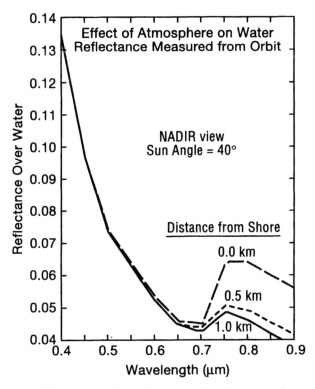

Figure 2-11. The impact of the adjacency effect for vegetation surrounding a lake.

being observed. As can be seen from the graph, even 1.0 km from shore there is a significant bump in the spectrum at 0.75 μm that is due to the vegetation reflectance. This shows quite graphically that, due to atmospheric scattering, the spectral response at a given point on the Earth's surface can be significantly affected by what is present on the Earth's surface at nearby (and not-so-nearby) points. This is referred to as the *adjacency effect*.

Clouds

49

Clouds pose a significant limitation to passive optical remote sensing systems. At first thought, one might tend to underestimate this limitation. One might tend to think of them in a binary fashion: they are either there or not. However, it is more realistic to think of the effect they have on the ability to use passive, optical means as a continuum of influences, varying from clear to haze to something easily identified by the unaided eye as a cloud. They affect both the illumination and the view of the surface; they vary in this effect spatially, spectrally, and temporally, and these variations are of high frequency. That is, one can expect the effect to vary significantly from pixel to pixel, from wavelength to wavelength, and from minute to minute.

The possible existence of cloud cover significantly affects the planning for data gathering in remote sensing. Clouds in their various forms significantly affect the measured pixel radiance even if they are not in a direct line of sight from the sensor to the pixel under consideration, for they can and do affect the illumination of the pixel. There high spatial variation makes it difficult to contemplate the adjustment of measured radiance data for atmospheric effects even in cases where it is thought desirable to do so. Their temporal nature has over time very much limited the usefulness of a satellite sensor with a typical two-week revisit time. Given the inability to control or change the satellite orbit, visitation to a desired site can be delayed by months in obtaining data over a desired area. Given the fact single satellites rather than a coordinated group of them has been the rule, this alone has greatly reduced the utilization of satellite sensors for operational purposes.

2.5 Sensor Optics

Next are examined some of the key parameters of the sensor system optics. A brief review of elementary optics may be useful at this

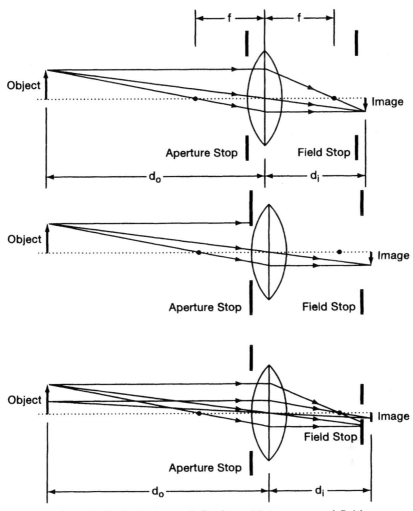

Figure 2-12. Optical stop definitions. (a) Aperture and field stops open. (b) Aperture stop closed down, fewer rays reach the image plane, so it will be darker (lower contrast). (c) Field stop closed down, portion of image cut off.

point, showing the effect of optical stops. Thin lens relationships used here assume that the lens focal length is much greater than lens thickness, a circumstance that is the case in remote sensing.

51

See Figure 2-12a. If d_o is the distance from the object to the lens, d_i is the distance from the lens to the image it is to form, and f is the distance from the lens to the focal point, called the focal length, It can be shown in this case that

$$\frac{1}{f} = \frac{1}{d_o} + \frac{1}{d_i}$$

Rays parallel to the longitudinal optical axis, passing through the lens are refracted through the focal point on the other side of the lens. Similarly, rays from an object passing through the focal point on the near side of the lens are refracted to be parallel to the optical axis on the other side of the lens.

From the figure it can be seen that if the aperture stop is closed down somewhat (Figure 2-12b), the image will be complete but darker. This is because some of the rays will be blocked by the aperture even though there are still rays from all parts of the object arriving at the image plane. If, on the other hand, the field stop is partially closed down (Figure 2-12c), the image will be bright but incomplete. Thus the aperture stop determines the light-gathering ability or *speed* of the lens. Note that as the aperture is closed down, the angular deflection of the rays at the edge of the aperture decreases, so this angular deflection is related to the light-gathering ability. Note also that as the focal length is shortened, the angular deflection also increases. Thus for two lenses with the same aperture but different focal length, the shorter focal length one has the greater light-gathering power.

A convenient way to express this mathematically is the f/D ratio. See Figure 2-13. The actual angle of deflection is referred to as the field of view and can be found from this figure using simple trigonometry:

$$\tan \beta \approx \beta = \frac{2d}{f}$$

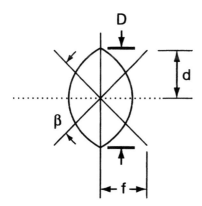

Figure 2-13. Angular relationships for a thin lens.

$$\beta = FOV = 2 \tan^{-1}\left(\frac{d}{f}\right)$$

For small fields of view, $\tan x \approx x$, and

Note from the thin lens formula above that for remote sensing, since d_o is usually very large, $f \approx d_i$.

Consider next the effect of field of view and instrument aperture in making an observation of a pixel. See Figure 2-14. The area observed by the instrument, ΔA, is equal to $(\beta H)^2$ where β is a plane angle in radians[3]. ΔA is defined by the field stop. Light from each single point within ΔA goes out in all directions, and the instrument aperture of area A_p, defined by the aperture stop, determines what portion of this light is collected by the lens. This aperture subtends a solid angle, $\Delta\omega$, defined by $\Delta\omega = A_p/H^2$. The larger the area being viewed and the larger the aperture being used, the greater the radiation received, thus it can be shown that the power into the sensor is proportional to $\Delta\omega \cdot \Delta A$, and this is equal to

3 β is often referred to in remote sensing as the Instantaneous Field of View, or IFOV.

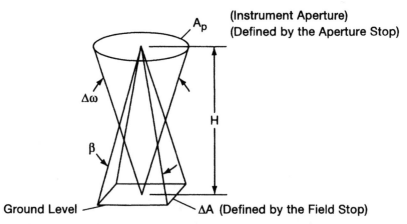

(Instrument Aperture)
(Defined by the Aperture Stop)

A_p

$\Delta\omega$

H

β

Ground Level

ΔA (Defined by the Field Stop)

Figure 2-14. The geometry of a sensor system observing a ground area.

$A_p\beta^2$. Note by the way, that this is independent of H (neglecting any effect of the atmosphere). These relationships will be used shortly in order to establish the signal-to-noise ratio for a sensor system.

2.6 Describing Surface Reflectance

A perfectly diffusing surface that reflects energy equally in all directions from an incoming ray is called a Lambertian surface. Natural surfaces are not Lambertian. Thus the reflectance of a surface depends on the direction from which it is illuminated and the direction from which it is viewed. See Figure 2-15 for nomenclature defining the illumination and view angles. There are several conventions for describing the reflectance of a surface. Three will be considered here.

<u>Bi-directional Reflectance Distribution Function (BRDF)</u>
Bi-directional reflectance distribution function is defined as the differential radiance leaving a point on the sample as a function of the viewing azimuthal and zenith angles, to the differential irradiance, as a function of its azimuthal and zenith angles.

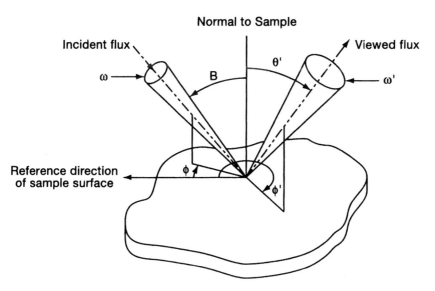

Figure 2-15. Geometric parameters describing reflection from a surface:
θ = zenith angle, ϕ = azimuthal angle, ω = beam solid angle; A prime on
a symbol refers to viewing (reflected) conditions.

$$f(\theta,\phi;\theta',\phi') = \frac{dL'(\theta',\phi')}{dE(\theta,\phi)}$$

dL' (W/(m²-sr) is the reflected radiance in the θ', ϕ' direction
produced by the incident irradiance dE (W/m²) of a collimated beam
from direction θ, ϕ. In both cases, θ refers to the angle from zenith
and ϕ to the azimuthal angle. The measurement of BRDF is quite
complex, since, though measurement of dL'$(\theta',\phi')$is
straightforward, dE(θ, ϕ) is inconvenient to measure. BRDF is
primarily of theoretical and developmental interest.

Bi-directional Reflectance
Bi-directional reflectance is defined as

55

$$d\rho(\theta, \phi; \theta', \phi') = \frac{dL'(\theta', \phi') \cos \theta' d\omega'}{L(\theta, \phi) \cos \theta d\omega}$$

The cos θ and $d\omega$ terms arise from the conversion from irradiance E (watts/m^2) to radiance L (watts/m^2-str). The dependence on θ and ω make it dependent on the instrument configuration used to make the measurement.

Bi-directional Reflectance Factor (BRF)
Normally so complete a description of the surface reflectance characteristics as BRDF is not needed for practical purposes. BRF is a more limited measure that is adequately complete but more easily measurable and not dependent on the measuring instrument. BRF is defined as the ratio of flux reflected by a target under specified irradiation conditions, to that reflected by an ideal completely reflecting, perfectly diffuse surface identically irradiated and viewed. Perfectly diffuse means reflecting equally in all directions. In practice, a smoked magnesium oxide or barium sulfate painted surface is sometimes used as a reference surface.

Figure 2-16 shows a field measurement system in use for measuring BRF. The reference surface is mounted above the roof of the cab of the elevated platform vehicle, such that the field spectrometer instrument head in the elevated platform can make a measurement of the subject material, vegetation in this case, then swing periodically over and measure the comparative response from the reference surface.

In effect, obtaining a quantitative description of how a surface reflects irradiance, while it may at first seem simple, is quite a difficult and complicated matter. Summarizing then, we have

Figure 2-16. An example field data collection system used for measuring BRDF.

defined three ways to characterize the reflectance characteristics of a scene element:

1. *Bi-directional Reflectance Distribution Function* provides a full description of the surface element but is impracticably complex.
2. *Bi-directional Reflectance* is a quantity often spoken of in practice, but one not unique to the scene element, it being dependent on the measurement instrument as well.
3. *Bi-directional Reflectance Factor* is one that while not providing a complete description of the scene element reflecting characteristics, it does provide a practically useful one and is not dependent upon the measuring instrument.

2.7 Radiation Detectors

There are several physical mechanisms that can be used for measuring optical power level quantitatively. Photographic film, a photochemical process, is one. However, it is best suited for making images rather than quantitative measures of optical power. Others that are more common for remote sensing purposes are thermal detectors, photon detectors, and photomultipliers. Each of these will next be briefly described.

Thermal Detectors

The principle involved here is that the incident radiation is absorbed and thus converted to heat. This causes a temperature change and therefore a change in electrical resistance. This resistance change is electrically measurable. An advantage of this method is that it is not dependent on wavelength and will thus work over a broad range of bandwidths. A disadvantage, however is that it is relatively slow due to thermal time constant of the sensor material.

Photon Detectors

The basic principal for photon detectors is quantum mechanical. Incoming photons excite electrons to a higher energy level where they are less bound to the atom nucleus and are free to move through the crystal lattice structure of the detector material; that is, they become charge carriers, raising the conductance of the material. Photon energy is given by

$$E = h\nu = \frac{hc}{\lambda}$$

where

h = Planck's constant
$\nu = c/\lambda$ = frequency
c = speed of light
λ = wavelength of the light

To activate the necessary increase in conductance, the photon energy must $\geq E_g$, the quantum difference between the valence band and the conduction band for the material. The sensor response is proportional to the number of charge carriers in the conduction band. **Responsivity** is a measure of the electrical response per watt of incoming radiation. Since

$$hc/\lambda \geq E_g \quad \text{joules}$$

$\lambda \leq hc/E_g = \lambda_c$, called the **cutoff wavelength.** The incident beam power, Φ:

$$\Phi = N_\Phi \frac{hc}{\lambda} \quad \text{watts}$$

where N_Φ is the number of photons/second. *Responsivity* of a detector, $R_\Phi = k_1 \eta N_\Phi$ is then

$$R_\Phi = \frac{k_1 \eta \Phi \lambda}{hc} \quad \text{volts/watt,} \quad \lambda \leq \lambda_c$$

where η is the quantum efficiency and k_1 the proportionality constant.

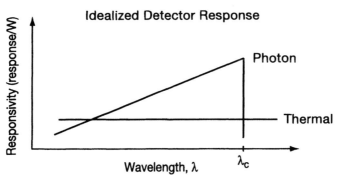

Figure 2-17. Idealized responsivity vs. wavelength for thermal and photon detectors.

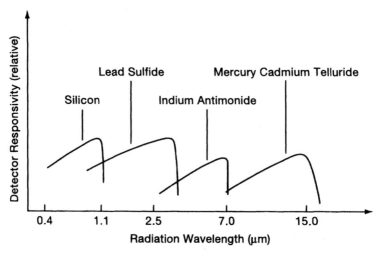

Figure 2-18. Measured relative responsivity wavelength ranges for several common photon detector materials.

Figure 2-17 shows an idealized graph of relative Responsivity plotted against wavelength for thermal and photon detectors.

One might need to choose a collection of detectors to cover the range of wavelengths of interest, each using sensitivity regions below the cutoff for that material but reasonably high on the curves. Example graphs of measured relative responsivity wavelength range for several common photon detector materials are given in Figure 2-18, showing how this could be done to cover a broad range of wavelengths. Note the similarity of the shape of these curves to the idealized form in Figure 2-17.

It is useful to have a means for comparing photon detector materials. Such a detector material figure of merit may be defined as follows. Necessarily there will be noise generated within a sensor. The Noise-Equivalent Power (NEP) is the power required to produce a signal just equal to the internal noise, $S/N = 1$. Now let $D = 1/NEP$. Since,

NEP is usually proportional to $\sqrt{\text{Area}}$ of the detector, a figure of merit for detector materials commonly used is

$$D^{*} = \frac{\sqrt{A}}{\text{NEP}}$$

D* also depends on the wavelength λ, f, the optical chopping frequency, and BW the bandwidth of the recording system. Handbooks give D* plotted against λ. Figure 2-19 shows the typical D* for several materials, some for the photoconductive mode and some for the photovoltaic mode.

Depending on the material, the signal can manifest itself cither as a change in resistance (**Photoconductive** - PC) or as an internally generated voltage (**Photovoltaic** - PV). To measure the effect of

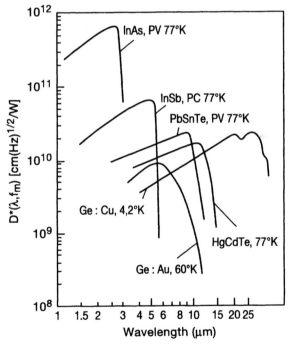

Figure 2-19. Figure of merit, D*, for several materials vs. wavelength.

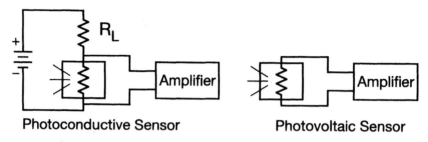

Photoconductive Sensor   Photovoltaic Sensor

Figure 2-20 Circuit arrangements for photoconductive and photovoltaic sensors.

the radiation on the sensor, a bias circuit is necessary in the photoconductive case. See Figure 2-20.

Bias circuitry noise generally is the limiting factor in terms of noise vis-'a-vis performance in the photoconductive case. Photovoltaic materials generally offer a higher S/N, but not many materials are photovoltaic. Johnson or thermal noise is often the predominant source of noise in sensors. It is white (meaning the same magnitude at all wavelengths) and zero mean Gaussian with mean square value given by

$$\overline{v^2} = 4kTR\Delta f$$

where k is Boltzmann's constant, T is the temperature of the resistor in °K, R is the resistance, and Δf is the bandwidth of the circuit. Sensor system noise will be treated more thoroughly in Chapter 9.

Another issue that arises for detector materials is cooling. The temperature of the detector must be sufficiently below the temperature of the effective radiation source so that internally generated noise is sufficiently low. This is usually easily achieved in the reflective portions of the spectrum since the effective temperature of the source (the Sun) is 6000 °K. However, for Earth sensing in the thermal region, it is necessary to cool the detectors to reduce their operating temperature well below the temperature of the target being observed ($\approx$ 300 °K).

Photomultipliers

The photomultiplier tube is another scheme that has been used to measure optical radiation. A diagram showing such a device schematically is given in Figure 2-21. The optical radiation is focused upon the photocathode, which emits electrons in proportion to the optical power level focused upon it. However, the key feature of the device is that it can amplify the signal through the use of the dynodes. In operation, each dynode is biased about 100 v positive to the previous one, so that the electrons are attracted to each in order with increased energy. Because of this, each dynode emits more electrons than arrive onto it. This enables an electron multiplicative effect. Photomultipliers are useful primarily in the visible and very near IR, $\lambda \leq 0.9$ μm. They were used on Landsat

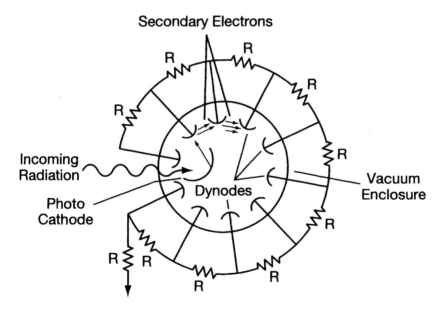

Figure 2-21. Schematic diagram of a photomultiplier detector.

MSS for the 0.5–0.6, 0.6–0.7, and 0.7–0.8 μm bands. Silicon detectors were used for the 0.8–1.1 μm band.

2.8 Sorting Radiation by Wavelength

Clearly, one of the important steps in a sensor system is the dividing of radiation arriving at the pupil of the sensor into the wavelength bands that are desired. There are several mechanisms that can be used for this purpose (see Figure 2-22). One is by use of a prism. In this case light passing through the prism is refracted through different angles depending on the wavelength of the radiation. This is a straightforward process in which detectors are placed at the proper location at the output to receive the radiation of the desired wavelength. However, there are several possible limitations of this method that may come into play. For one, the angle of deflection is limited so that it may difficult physically to divide the radiation

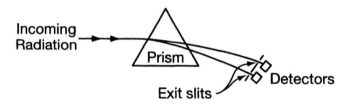

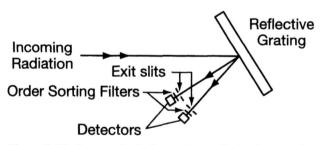

Figure 2-22. Two methods for sorting radiation by wavelength.

into a large number of narrow bands. It is also the case that prisms made of a given material may not be effective over the entire desired range of wavelengths.

A second possible means is via a reflective or partially transparent grating. In this case the discontinuities on the grating surface result in the radiation being redirected at an angle dependent on the wavelength of the radiation. A characteristic of this process is that frequencies (= c/λ) of integer multiples of each other are redirected in the same direction. Thus it is necessary to use order-sorting filters to choose only the radiation of the frequency of interest.

Gratings are generally able to spread the radiation over a wider area than are prisms. Thus they may be more easily used to divide radiation into a larger number of more narrow wavelength bands.

A third mechanism for wavelength sorting is to use constructive or destructive interference. The idea is to divide the radiation into two paths, one of slightly different length than the other, so that there is a phase difference at a given wavelength when they are brought back together. Then, as shown in Figure 2-23, wavelengths that are such that they are in phase will reinforce one another, while those out of phase will cancel each other out. The phase will depend on the difference of the lengths of the paths the two have passed through.

This type of mechanism can be used to form a filter by making dielectric layers that are partially reflective and partially transmissive, as shown in Figure 2-24. If the layers are made just half as thick as the wavelength it is desired to pass, the filter will effectively pass just that wavelength.

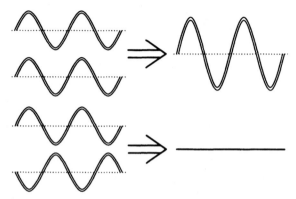

Figure 2-23. Constructive and destructive interference of two wavelengths of the same frequency.

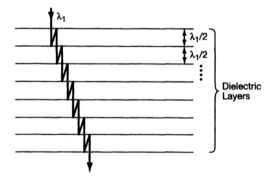

Figure 2-24. The construction of a wavelength filter using constructive interference for the desire wavelength.

2.9 Multispectral Sensor Systems

Putting the needed optics, detectors, and related components together in a practical sensor system has evolved significantly over the years as the technology has developed. In this section the chronology of this development will be briefly outlined so that a perspective can be gained about how the technology arrived at its current state and what the controlling factors are in that continuing process. Rather than attempting to describe specific current systems, which are rapidly superceded by newer ones, we will focus on the fundamental factors involved in this continuing development.

The earliest systems took the form of line scanners. The basic idea of a line scanner is to use some type of scanning mechanism to scan a spectrometer's instantaneous field of view (IFOV) across the scanner's total field of view (FOV), transverse to the direction of platform travel. The platform travel provides the motion in the perpendicular direction so that a two dimensional region is scanned in time.

Figure 2-25 shows a very simple example mechanism. The collecting optics define the system aperture A_p. The field stop (usually the detector) defines the pixel size or IFOV. Various methods have been used to define the wavelength range of each spectral band. In the case shown, the front surface of the element marked dichroic grating diverts some of the energy to a prism to

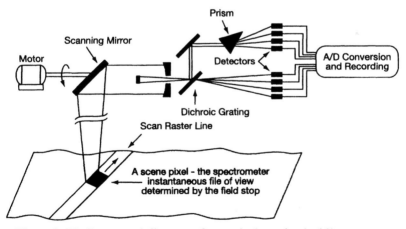

Figure 2-25. Conceptual diagram of an optical-mechanical line scanner as might be mounted in an aircraft.

disperse energy for some of the wavelengths, while for other wavelengths the remaining energy passes though the dichroic grating element to be wavelength dispersed by a grating on the back side of the element. Detectors for each wavelength band then convert the optical power to an electrical signal that is then digitized and recorded.

A very convenient characteristic of this simple arrangement is that for a given spectral band a single detector is used for every pixel in the scene. This makes the data inherently consistent across the scene, and calibration of the spectral band is a simpler process.

The disadvantage of the arrangement is that for a given platform velocity and FOV, less time is available to dwell on or measure the power level of a given pixel area, thus affecting the signal-to-noise ratio. This turns out to be a key limitation that had to somehow be mitigated as the sensor technology was moved from aircraft to the much higher velocity early spacecraft systems.

Spectral Resolution, Spatial Resolution, and S/N

Given the law of conservation of energy, these three–spectral resolution, spatial resolution, and S/N– become the fundamental parameters of the data collection process These parameters relate how the energy coming up from the scene is divided up spectrally and spatially and what power level is left over to overcome the noise inherent in the sensor system. S/N then defines how precisely the power level of a pixel can be quantified. Next will be developed the relationships amongthese three fundamental parameters.

For a given spectral band, the power into the instrument from the surface in the interval from λ to $\lambda + \Delta\lambda$ is

$$\Phi = \tau_a \, L_\lambda \, A_p \, \beta^2 \, \Delta\lambda \qquad \text{watts}$$

where τ_a = atmospheric transmittance

L_λ = spectral radiance of ground pixel, W/(m²-sr-μm)

A_p = scanner aperture, m²

β = scanner IFOV, radians

Then

$$L_\lambda = E_\lambda \, R \cos \theta_s$$

where

E_λ = spectral irradiance on the ground pixel, W/(m²-μm)

θ_s = sun angle, radians

R = bidirectional reflectance factor, dimensionless

Sensor noise serves to limit the degree to which two different reflectance levels can be distinguished. Consider the situation of Figure 2-26. The change in power level, $\Delta\Phi$, for example, in a 10% field to a 11% field, is given by

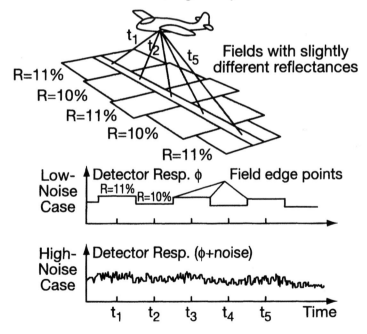

Figure 2-26. Noise limitations on reflectance resolution.

$$\Delta\Phi = \tau_a \Delta L_\lambda A\beta^2 \Delta\lambda = \frac{\Delta L_\lambda}{L_\lambda}\tau_a L_\lambda A\beta^2 \Delta\lambda \quad watts$$

Therefore

$$\Delta\Phi = \frac{\Delta L_\lambda}{L_\lambda}\Phi$$

In a similar manner it may be shown that

$$\frac{\Delta L_\lambda}{L_\lambda} = \frac{\Delta R}{R}$$

Thus

$$(\Delta\Phi)_{refl} = \frac{\Delta R}{R}\Phi \quad watts$$

For some typical values, if

$$\tau_a = 0.5$$
$$L_\lambda = 28 \ W/(m^2\text{-}\mu m\text{-}sr)$$
$$A_p\beta^2 = 5.37 \times 10^{-8} \ m^2\text{-}sr$$
$$\Delta\lambda = 0.68\text{--}0.62 \ \mu m = 0.06 \ \mu m$$

Then

$$\Phi = 4.5 \times 10^{-8} \ watts$$

And for a 1% change in reflectance at 10%,

$$\Delta\Phi = 4.5 \times 10^{-9} \ watts$$

Now the scan rate across the field of view must be coordinated with the platform's forward velocity so that the platform moves just one pixel width during the time of one scan line. One of the ways to add flexibility to that process is to have a scan mirror with more than one active side so that for each scan motor rotation more than one scan line is covered. Also, instead of using one detector for a given spectral band, several detectors could be arranged so that several scan lines are measured at once. See Figure 2-27.

Assume a scanner with the following variables:

 p - sided scan mirror
 q - detectors in an array
 f - primary focal length
 $(\pi/4)D^2$ - effective primary mirror area
 ω - angular rotation rate - radians/sec.
 β - scanner IFOV, radians
 H - scanner altitude
 V - platform velocity

Then the time interval required to move one pixel width is β/ω. The rise time of the electronic amplifier/recorder, τ, must be short

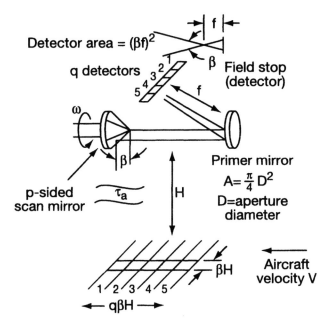

Figure 2-27. A multisided scan mirror using more than one detector.

compared to the time to sweep one pixel width. Let g be factor relating these quantities:

$$\tau = \frac{1}{g}\frac{\beta}{\omega} \quad \text{seconds}$$

There are q detectors viewing q pixels at once in the platform velocity direction. The width of a pixel is βH, and the along flight distance swept out is $q\beta H$. The full width of the area scanned is covered in $2\pi/p$ radians of scanner rotation. The scan mirror must rotate $2\pi/p$ radians in the time it takes to advance $q\beta H$ meters, meaning in $q\beta H/V$ seconds. Thus the mirror must spin at,

$$\omega = \frac{2\pi V / H}{pq\beta}$$

Then the electronic circuit rise time must be,

$$\tau = \frac{\beta}{g\omega} = \frac{pq\beta^2}{2\pi g V / H} \quad \text{seconds}$$

The electronic circuit bandwidth, BW, is related to rise time by $BW = 1/a\tau$ hertz, where a is usually between 2 and 3. Thus

$$BW = \frac{2\pi g V / H}{apq\beta^2} \quad \text{Hz}$$

The noise-equivalent power for a detector was previously determined as

$$NEP = \frac{\sqrt{A * BW}}{D_\lambda^*} = \frac{1}{D_\lambda^*}\sqrt{\frac{f^2\beta^2 2\pi g V / H}{apq\beta^2}} = \frac{1}{D_\lambda^*}\sqrt{\frac{2\pi g V / H}{apq}} f \quad \text{watts}$$

The optical efficiency, η, must also be taken into account. The input signal power is

$$\Delta\Phi_\lambda = \tau_a \Delta L_\lambda \frac{\pi}{4} D^2 \beta^2 \Delta\lambda\eta \quad watts$$

Divide the last two equations to obtain the signal-to-noise ratio.

$$\frac{\Delta\Phi_\lambda}{NEP} = \frac{\sqrt{\pi}\tau_a\Delta L_\lambda DD^*_\lambda\sqrt{apq}\beta^2\Delta\lambda\eta}{4\sqrt{2g}\sqrt{V/H}f/D}$$

To simplify, let $p = 4$, $a = 3$, $g = 1.5\pi$, and $\eta = 0.4$. Then the equation reduces to,

$$\frac{\Delta\Phi_\lambda}{NEP} = \Delta L_\lambda \frac{DD^*_\lambda\sqrt{q}}{20\sqrt{V/H}f/D}\beta^2\Delta\lambda$$

Notice that, this signal-to-noise ratio increases with,

- Scanner aperture, D
- Speed of the optical system, D/f
- Detectivity D^*_λ
- Number of detectors, q
- Field of view, β, namely the angular pixel size
- Spectral bandwidth

S/N decreases with the V/H ratio, that is, as the platform velocity increases or the altitude decreases, but only in a square root fashion. Notice, though, that both spectral resolution, $\Delta\lambda$, and spatial resolution, β, come at a cost, especially spatial resolution. And finally, attention has previously been drawn to the fact that from a broad systems viewpoint, the fundamental parameters for obtaining information about an area remotely are

- The spectral resolution
- The spatial resolution
- The signal-to-noise ratio

Notice that the S/N relationship above shows how the three interrelate, that is, the S/N varies linearly with the spectral resolution, $\Delta\lambda$, and as the square of the spatial resolution, β. The first term on the right shows that the S/N varies linearly with illumination, ΔL_λ, and the fraction on the right shows the influence of the key sensor system parameters.

Analog to Digital Conversion

Given the electrical signal emerging from the detector, the next step in the process would be conversion of the analog signal to discrete digital values. Figure 2-28 gives a simple illustration of how this process works. The signal for each band is sampled at the appropriate instants, and then for a given sample, the signal must be converted to discrete binary numbers. This is done by sorting the magnitude at the particular sampling instant into increasingly fine binary values within the overall dynamic range. The overall dynamic range must be set by the system designer or operator by deciding what are the brightest and the darkest values that will be encountered. Then at a given sampling instant, the most significant bit indicates whether the signal level at that instant is in the upper half or lower half of that maximum dynamic range. Subsequent bits indicate whether the signal level is in the upper or lower half of each subsequent magnitude range, as shown in Figure 2-28. For example, for the 3-bit system of the figure, there are a total of 2^3 or 8 possible values, namely, levels 0 through 7. The value at time t_2 would be recorded as 011, because the signal magnitude is in the

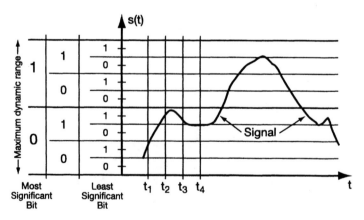

Figure 2-28. An analog signal being sampled at times t_n and quantized to discrete digital values.

lower half of the most significant bit range and in the upper half of the middle and least significant bit ranges. The binary number 011 (=0*4+1*2+1*1) indicates level 3 of the 0–7 possible values.

Early aircraft systems typically had 8-bit data systems indicating 256 possible shades of gray for each pixel in each band. Landsat MSS, the first spaceborne system had a 6-bit data system or 64 shades of gray for each pixel and band. Modern sensor systems typically have 10- or 12-bit systems indicating 1024 or 4096 possible gray values based on the much higher S/N ratios obtainable by current sensor systems. However, it is important to keep in mind that a data set received from a 10-bit sensor data system will very likely have very much less than 10 bits (1024 shades of gray) active in the data set. Setting the sensor dynamic range for the brightest and darkest values anticipated over the usage time of the sensor may mean that in any given data set a much smaller range of binary values will be active if the particular area covered in the given data set has no pixels approaching those extreme brightness levels. Further, usually system designers and operators set the limits of the maximum dynamic range quite conservatively to ensure that saturation does not occur.

Geometric Considerations

To further complicate matters, there are some geometric factors to be taken into account in the data collection process. Figure 2-29 indicates some key ones regarding an airborne or spaceborne scanner are

X - (down-track) Direction of flight
Y - (cross-track) Scanning direction
Z - Altitude (above the reference plane) direction
β - Angular resolution in the down-track direction - in radians
γ - Angular resolution in the cross-track direction - in radians
θ - Angle of scan deflection

α - Maximum scanner deflection angle, i.e., $|\theta| \leq \alpha$
Z_c - Altitude of the scanner above the reference plane
Z_j - Altitude of the jth pixel above the reference plane

Note that the following kinds of distortion occur even in the ideal case of a perfectly stable platform:

- If sampling at equal increments of θ is assumed, pixels will grow longer and wider as θ increases.
- Pixels will become centered further apart on the ground.
- Pixel area as well as the degree of underlap or overlap of pixels will change with θ in both the along-track and the down-track directions.

These effects are known as *scan angle effects*.

- Note also that as the height of terrain above the reference plane increases, the pixels will grow smaller, and conversely.

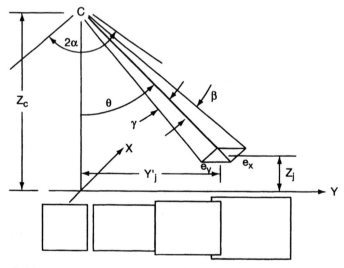

Figure 2-29. Geometric parameters of scanning showing the growth of pixel size and overlap as θ increases.

- Note that the area of the ground covered by a pixel will change not only with the elevation of the pixel, but also with its slope and aspect.

These are referred to as *topographic effects.*

- Note that if scanning in the Y direction is not instantaneous, and there is motion in the X direction continuously, the location of a row of pixels on the ground is not in a straight line and not perpendicular to the X direction.

This is referred to as *scan time effect.* These effects, especially the former two, tend to be more serious at aircraft altitudes than they are at satellite altitude.

In practice, additional distortions are introduced due to the fact that the platform will have roll, pitch, and yaw components of motion as well. Figures 2-30 and 2-31 illustrate these distortions.

Sun Angle Effects

And finally, the question of the sun angle must be taken into account. For spacecraft, this has to do with the particular time of day, season and latitude chosen for the data collection. Orbits for land oriented spacecraft are usually chosen as sun synchronous, near polar ones. By this is meant an orbit that tends to pass over a given site at the same local time for that site. This is accomplished by choosing an orbit that passes over both poles of the Earth and whose orbital plane has a fixed orientation with regard to the sun-Earth line. Figure 2-32 illustrates this for a 9:30 AM local passage time. The choice of local passage time usually is selected based on a compromise between achieving the highest solar zenith angle, and therefore the brightest illumination and thus the highest S/N ratio, and the time of minimum likely cloud buildup. In the mid-latitude growing season, atmospheres tend to be the clearest in the morning with cumulus buildup as the day progresses. Thus 9:30 to 10:30 AM has been a typical compromise passage time.

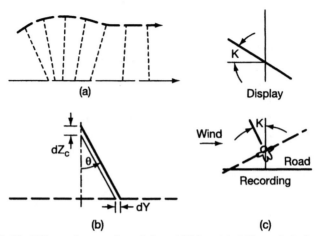

Figure 2-30. Effect of some aircraft instabilities. (a) Effect of pitch, (b) Effect of flying height, (c) Effect of yaw.

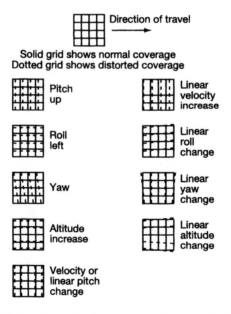

Figure 2-31. Resultant Earth coverage of scanner in the presence of different vehicle orientation irregularities.

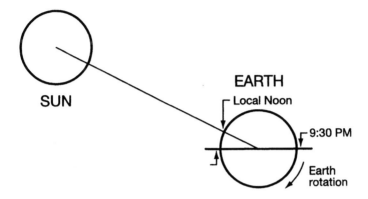

Figure 2-32. The Earth-Sun relationship showing a common choice for a land satellite orbit.

2.10 The Development of Multispectral Sensor Systems

The multispectral concept and sensor systems for collecting multispectral data from aircraft and space developed steadily over the last third of the 20th century. It is useful to know something of this development in order to see how information systems based on them emerged from being sensor limited to its current circumstance. In closing this chapter, specifications for a few historically significant example systems follow to illustrate how sensor systems have evolved.

M-7 airborne system

The M-7 airborne scanner system[4] was one of the earliest systems used to develop land multispectral technology. It was built and operated, beginning in the early 1960's, by the University of Michigan Willow Run Laboratories, which later became the Environmental Research Institute of Michigan (ERIM). It was initially flown on a C-47 aircraft. Its overall concept was similar to that in Figure 2-25, using a rotating scan mirror scanning a single row of pixels at a time.

The M-7 airborne scanner was operated with various spectral band configurations, one of which is indicated in Table 2-5. This configuration was used for the 1971 Corn Blight Watch Experiment[5]. This was a survey of sample segments over the entire western third of the state of Indiana every two weeks of the 1971 growing season. The purpose was to monitor the spread of a particular strain of corn blight over that season. This experiment was one of the first major tests of an operational nature of the newly emerging multispectral technology, and it served to benchmark progress in developing the technology to that point. It demonstrated that not only could an important agricultural crop specie be identified and mapped with multispectral data, but the degree of infestation of a particular pathogen in that specie could be at least crudely measured, and this could be done over a large geographic area and over an entire growing season.

---

4 P.H. Swain and S. M. Davis, eds., *Remote Sensing: The Quantitative Approach*, McGraw-Hill, 1978, p. 122.

5 R. B. MacDonald, M. E. Bauer, R. D. Allen, J. W. Clifton, J. D. Ericson, and D. A. Landgrebe, Results of the 1971 Corn Blight Watch Experiment, *Proceedings of the Eighth International Symposium on Remote Sensing of Environment*, Environmental Research Institute of Michigan, Ann Arbor, Mich. pp 157-190, 1972

Band Number	$\Delta\lambda$ in μm	Band Number	$\Delta\lambda$ in μm
1	0.40–0.44	10	0.66–0.72
2	0.44–0.46	11	0.72–0.80
3	0.46–0.48	12	0.80–1.0
4	0.48–0.50	13	1.0–1.4
5	0.50–0.52	14	1.5–1.8
6	0.52–0.55	15	2.0–2.6
7	0.55–0.58	16	4.5–5.5
8	0.58–0.62	17	9.3–11.1
9	0.62–0.66		

Table 2-5. Spectral band configuration of the M-7 optical mechanical scanner as used for the 1971 Corn Blight Watch Experiment.

Additional system specifications of the M-7 system are given in Table 2-6. A/D conversion was carried out on the analog data tapes on the ground. This conversion was to 8 bit precision, meaning that there were $2^8 = 256$ possible gray values for each pixel in each spectral band. It was a key job of the system operator on board the aircraft and the A/D converter technician on the ground to set the system amplifier gains for each band so that the lightest pixel in the area overflown did not exceed that dynamic range, causing a saturation condition.

- FOV = 90° (± 45° from nadir)
- Scan lines scanned per scanning mirror rotation, p = 1 (see Figure 2-27)
- No. of scanlines scanned at once, q = 1
- IFOV, β = 3 milliradian (nominal)
- Aircraft velocity, V = 60 m/s (nominal)
- BWa = 1/3
- Ratio of time to scan 1 pixel width to amplifier rise time, g = 1.5π (assumed)
- Atmospheric transmission efficiency, τ_a = 0.8 (assumed)
- 0.1 °C nominal thermal resolution
- 1% nominal reflectance resolution
- 12.25 cm diameter collector optics
- 60 or 100 scan/sec
- DC to 90 K Hz electronic bandwidth
- Roll-stabilized data
- Data recorded on board in analog form

Table 2-6. Specifications for the M-7 Airborne Scanner

Figure 2-33 shows the spectral bands of this configuration in relationship to the spectral reflectance curve typical of green vegetation. Notice how the bands of the visible region are quite narrow, but as the wavelength of the bands move out into the reflective, middle, and thermal IR, the bands are widened. This is due at least in part to the lower level of solar illumination there. See Figure 2-5. This makes it possible to maintain a reasonable S/N ratio there, but at the loss of some detail as a result. Nevertheless, it is seen that all major parts of the vegetative spectrum are represented in data from this system.

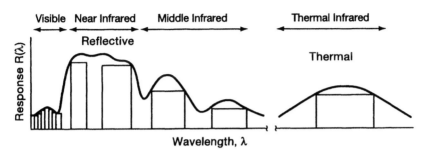

Figure 2-33. Location of the spectral bands of the M-7 system relative to the nominal spectral reflectance of green vegetation.

The configuration that was used for maintaining an internally consistent calibration is shown in Figure 2-34. The scheme used two calibration sources for both the reflective and the thermal region so that each calibration source was observed on each rotation of the scanner, namely, during each scan line. In the reflective region,

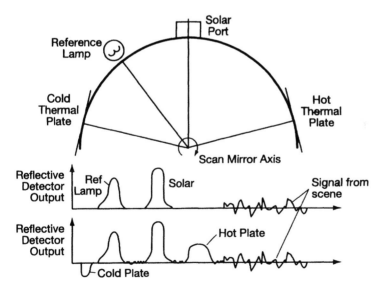

Figure 2-34. M-7 Scanner Calibration Sources and typical signal outputs for one rotation of the scanner.

83

originally this consisted of two calibration lamps, one at the darker and the other at the lighter end of the expected dynamic range. The most important job of the calibration system is to maintain internal consistency, namely, to ensure that a given digital count out of the A/D system always reflects the same level of radiation into the pupil of the sensor input. One of the lamps was later replaced by a fiber optics bundle leading to a port on top of the aircraft, in an attempt at calibration based on the solar illumination at any given time. However, this did not prove very helpful.

Landsat Multispectral Scanner (MSS)

By the late 1960s enough had been learned about such spectral technology that it was time to design a spaceborne version of such a sensor. This sensor was to be a companion of an image oriented one that was to use return beam vidicon tubes. The multispectral system came to be called simply Multispectral Scanner (MSS). However, space technology being rather young at that point, it was not felt possible to have more than four spectral bands in this first system (see Table 2-7). Further, since angular momentum is rather difficult to deal with on an orbiting spacecraft, the scanning mechanism was based on an oscillating mirror. Data were collected on the forward mirror oscillation, followed by a return motion to prepare for the next forward oscillation for the next scan line.

The much higher orbital velocity meant that to attain a reasonable S/N for the spatial and spectral resolution selected, $q = 6$ was needed, meaning that six scan lines needed to be swept out with each forward motion of the scan mirror. This, in effect, increased the dwell time on each pixel, allowing for the gathering of more energy from the pixel and thus the increase in S/N. Using $q = 6$ increased the complexity of the calibration process. However, as all six channels of each spectral band needed to have gains as precisely the same as possible.

Three such sensor systems were built and then launched in sequence in 1972, 1975, and 1978. Table 2-8 shows some of the major parametes of the MSS system. All three were essentially identical, except that the Landsat 3 MSS contained a fifth band in the thermal region.

Band Number	Spectral Band μm	Radiance for full output mW/(cm²-sr)	Detector Type	IFOVm	Dynamic Range-Bits
1	0.5-0.6	2.48	PM	79	7
2	0.6-0.7	2.00	PM	79	7
3	0.7-0.8	1.76	PM	79	7
4	0.8-1.1	4.60	Si	79	6
5[6]	10.4-12.6			237	

Table 2-7. MSS spectral bands.

- Orbital velocity V = 6.47 km/s
- IFOV = 0.086 mrad
- BW = 42.3 k Hz per band
- Optical efficiency = 0.26
- NEP = 11 x 10^{-14} W/Hz
- Entrance aperture = 22.82 cm
- Nominal orbital altitude H = 915 km
- No. of scanlines scanned at once, q =6
- Swath width = 185 km (FOV =11.56°)

Table 2-8. Landsat MSS Specifications

---

[6] Band 5 was carried on Landsat 3 only.

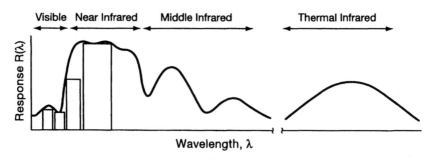

Figure 2-35. Location of the spectral bands of the MSS system relative to the
nominal spectral reflectance of green vegetation.

Figure 2-35 shows the location of the MSS spectral bands in relation
to a nominal green vegetation spectral reflectance. Clearly, these
four, broad, closely located bands do not model the spectral response
of green vegetation nearly as well as those of the M-7 bands. Thus
the young state of space sensor technology in 1968 imposed a
significant limitation on what could be done with Landsat data on
vegetation as well as all other types of problems. Nevertheless,
MSS proved to be productive for many tasks.

Thematic Mapper on Landsat 4, 5, 6[7], 7[8]
Given the rapid early success of MSS and despite its limitations,
by 1975 it was time to define a second-generation spaceborne
system. By this time the technology for constructing and operating
spacecraft sensors in orbit had advanced significantly, making
possible a sensor system with seven spectral bands at a finer spatial
resolution and a S/N ratio that justified an 8 bit data system. A
significant part of the reason for this was a design of the scanning
system, which, while still an oscillating mirror arrangement, could

---

[7] Landsat 6 TM referred to as ETM added a Panchromatic band
(0.50-0.90 μm) at 15 m. The 1993 launch failed.

[8] Landsat 7 TM referred to as ETM+ due to enhance calibration
capabilities, etc. Launched April 15, 1999.

collect data on both the forward and the return oscillation. A value of q = 16 was also used. This allowed for a substantially longer dwell time on each pixel, and thus an improved S/N even for the narrower bands and finer spatial resolution. Table 2-9 provides the spectral band delineation of Thematic Mapper. Figure 2-36 shows the bands again in relation to a representative green vegetation spectrum. Athough still not as thorough a sampling of the key wavelength regions as M-7 had been some years before, this was a substantial advancement over MSS.

Band Number	Spectral Band-μm	Detector Type	IFOV m	Dynamic Range in Bits
1	0.45-0.52	Si	30x30	8
2	0.52-0.60	Si	30x30	8
3	0.63-0.69	Si	30x30	8
4	0.76-0.90	Si	30x30	8
5	1.55-1.75	Si	30x30	8
7	2.08-2.35	InSb	30x30	8
6	10.4-12.5	HgCdTe	120-120[9]	8

Table 2-9. Thematic Mapper spectral bands.

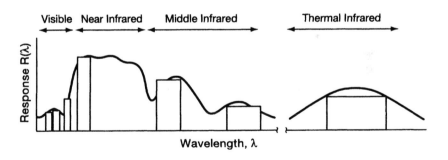

Figure 2-36. Location of the spectral bands of the Thematic Mapper system relative to the nominal spectral reflectance of green vegetation.

---

9   The thermal band on ETM[+] was improved to 60x60 m

Landsats 4, 5, 6, and 7 all carried Thematic Mapper instruments, with only slight modifications, from Landsat 4 launched in 1982 to Landsat 7 launched in 1999. The Landsat 7 instrument, called ETM+, has the same 185 km (±7.5° FOV) swath from 705 km sun-synchronous orbit as earlier versions, giving full Earth coverage in 16 days. One Thematic Mapper scene contains 5968 lines by about 7000 columns. This means almost 40 Mpixels per scene or about 292 Mbytes ($\approx$ 2.34 Gbits). The data rates and the data volume begins to become a problem as the number of spectral bands but especially the spatial resolution goes up. There is on board solid state memory for 375 Gbits (40 minutes of data at end-of-life). The data downlink is via three pointable X-band antennas, each at 150 Mbps.

The Hyperspectral Era

Continued advancements in solid state device technology, among other things, have in recent years made possible the next large advancement in sensor systems. These new methods and techniques make possible sensor systems with as many as several hundred bands over the visible/near IR/middle IR regions and with bands as narrow as 5 to 10 nm. One of the first airborne systems of this type was the NASA Airborne Visible/Infrared Imaging Spectrometer (AVIRIS) system, which was first flown in 1987. It provides data in 210 bands in the 0.4 to 2.4 μm region. Data from this and similar instruments is sometimes referred to as "imaging spectrometer data," paralleling the concepts used by laboratory chemical spectroscopists, because the fine spectral detail of the data can reveal molecular absorption bands, thus enabling individual minerals to be identified where pixels dominated by a given mineral is expressed on the surface. However, more generally, sensors with large numbers of bands are referred to as hyperspectral sensors, allowing for a broader range of data analysis concepts including those based on chemical spectroscopy concepts as a subset.

A number of implementations of such sensor systems for both airborne and spaceborne platforms have emerged. The methods used to achieve such high spectral detail while maintaining good spatial resolution and S/N vary. Some are simply extensions of those used in earlier sensors. Other possibilities include the use of two dimensional solid state arrays, where one dimension is used for the pixels in a scan line, while the other functions in the spectral dimension. Thus, since all pixels in a scan row are in view at the same time, for a given down track velocity, the dwell time on each pixel is in effect increased by a factor equal to the number of pixels in a row.

However implemented, the ability to achieve these degrees of spectral detail finally essentially eliminates the constraint of spectral detail as being the primary limitation on the ability to extract information from data on a spectral basis. At the same time the availability of such high-dimensional data moves the key information extraction limitation to the information extraction methodology arena. This is the topic to be pursued in the remainder of this book.

2.11 Summary

The essential elements of sensor systems have been briefly studied, including

- Radiance and spectral nomenclature,

- The sun, its spectral characteristics, and Planck's Law,

- Atmospheric effects on radiant power,

- Optics, lenses, and optical stops,

- Means for characterizing the reflectance properties of surfaces,

- Thermal, photon, and photomultiplier detectors, and

- Means for sorting radiation by wavelength.

Next, a discussion was given as to how these elements are assembled into sensor systems. The following equation, showing the interrelationship of key parameters was arrived at.

$$S/N = \frac{\Delta\Phi_\lambda}{NEP} = \Delta L_\lambda \frac{DD^*_\lambda\sqrt{q}}{20\sqrt{V/H}f/D}\beta^2\Delta\lambda$$

This equation shows the relationship between S/N, illumination level ΔL_λ, major sensor system parameters, the spatial resolution β, and spectral resolution, $\Delta\lambda$.

The chapter was concluded with a brief description of a progression of several historically significant sensor systems. From this can be observed how the S/N, β, and $\Delta\lambda$, these being the key parameters from the data analyst's point of view, were steadily improved over time. Chapter 3 will then begin the study of how data produced by such systems may be analyzed.

3 Pattern Recognition in Remote Sensing

In this chapter pattern recognition methods are explored in
detail as they apply to multispectral remote sensing data.
Particular attention is given to matters that limit performance
in terms of accuracy of classification. Both supervised and
unsupervised classification are discussed, as well as means
for measuring and for projecting the accuracy of results. The
chapter is concluded with a discussion of the steps needed in
analyzing a data set, and a specific example is presented.

3.1 The Synoptic View and the Volume of Data

The original motivation for using airborne and spaceborne sensors
to acquire information had to do with the synoptic view from
altitude. If one goes to higher altitudes, one can see more, and that
presumably would lead to a more economical way to gather the
data. Though initial costs in the system would increase with
altitude, the economies of scale were thought to increase faster.
But going to higher altitudes to see more also means greater
quantities of data must be dealt with, thus it is important to find
economical ways of analyzing the data to extract the desired
information if the needed economy is to be maintained.

As discussed in Chapter 1, it is natural to think of images when
contemplating such a problem, and perhaps the first parameter
that comes to mind when thinking of images is resolution, namely
spatial resolution. Indeed, the spatial resolution needed has much
to do with the cost of the sensor system, its operation, and the cost
of processing the data that it collects. Speaking very broadly, the
image processing world dichotomizes into two classes of problems,
according to the character of the subject matter in the images. One,
which might best be called picture processing, deals with scenes
that have a spatial-resolution-to-object-size ratio such that physical
objects are easily recognizable by the human vision system based

upon the scene spatial characteristics. See Figure 3-1. Images such as in Figure 3-1*a* and even those taken with a handheld camera are examples of "Pictures" in this sense, because they are ordinarily intended for human view. One may speak of such images as "being of human scale."

Figure 3-1a. A "Picture." Pixels are a few centimeters in size. Objects are readily identifiable by human view. Viewing the area of Figure 3-1b at this resolution would result in about 10^6 times as much data. (In color on CD)

Figure 3-1b. A Landsat image. Pixels are 30 meters in size. Objects are less identifiable by human view, but the volume of data is much more manageable. (In color on CD)

The second class of images has a spatial-resolution-to-object-size ratio such that objects are not easily recognizable by the human vision system. However, this does not necessarily have much to do with the amount of information derivable from the data. An image from the Landsat series of satellites is an example of this type of image. See Figure 3-1*b*. If the spatial resolution in such an image were increased by two orders of magnitude or so, individual trees in a forest could be seen and recognized by a human, but at Landsat resolution, if one can only view the image as a "picture," it would be difficult to reach that conclusion. Other means must be available to identify individual pixels of the image as forest. It is this possibility that we wish to study.

The point is, if information about large areas is needed, the latter class of data could be much more economically dealt with if ways can be found to derive the desired information from it. The spatial resolution being much lower means that the volume of data is much smaller, and, indeed, lower resolution sensors are much less expensive, being lighter in weight and requiring less precise pointing and a simpler attitude control systems. Thus a key to being able to take maximum advantage of the synoptic view from space *economically* is to use the lowest resolution that will provide the desired information. This is what led to the investigation of the use of spectral characteristics, as compared to spatial ones, and, indeed, to the possibility of labeling (i.e., identifying the contents of) individual pixels from a scene. The means to do so is contained within the spectral distribution of energy from each pixel, and pattern recognition methods are well suited to use as a means for deriving labels for each pixel.

3.2 What is a pattern
Pattern recognition as a means for analyzing data has been under development for many decades, and it really came into its own as the digital computer began to be widely available in the 1950s and

1960s. One may first think of pattern recognition as a way to recognize a geometric pattern. For example, it is the means by which optical character readers are able to recognize geometric patterns of black lines on a white background as specific alphanumeric characters. However, it is really a more general concept than that, for example, as it is also used for such tasks as recognizing words in spoken sounds, or objects in pictures. Actually, the word pattern in pattern recognition really refers to a pattern in a set of numbers.

The general scheme of a pattern recognition device is shown in Figure 3-2. A natural pattern is viewed or measured with some type of a Receptor containing a sensor. The Receptor reports out the measurements as a set of numbers. This set of numbers $X = [x_1, x_2, ... x_n]$, usually represented as a vector, one for each pixel of the scene, is then fed to a Classifier which makes a decision as to what pattern the set of numbers of each pixel represent. Next we will explore how to use this concept for the multispectral remote sensing problem.

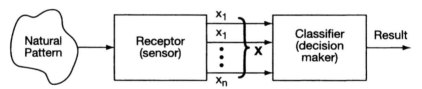

Figure 3-2. Diagram of a pattern recognition system.

The Receptor: A Transformer of Spectra
Multispectral remote sensing begins with the hypothesis that

- The spectral distribution of energy as a function of wavelength rising from a pixel somehow characterizes what Earth surface cover type the pixel contains.

To investigate this concept, consider the following. Shown in Figure 3-3a on the left is the result of carefully measuring in the laboratory and plotting the spectral response of a particular soil type at 200 different wavelengths. Assuming the above hypothesis, one might at first think that such a curve fully characterizes that soil type.

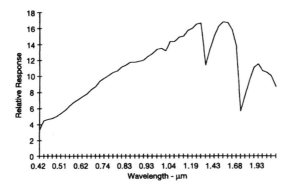

(a) Result of a single measurement

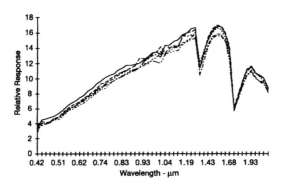

(b) Result of multiple measurements.

Figure 3-3. Spectral response of a specific soil type vs. wavelength.

However, the spectral response must be examined more fully. Shown in Figure 3-3b is the result of measuring the same soil sample at five slightly different locations. One notes that the five curves are not exactly identical, though they are similar. It might at first be assumed that the variation exhibited is noise, but this is not the case. The variations are stochastic in nature, but they are information-bearing, not noise. Soil, like any other Earth surface cover, is a mixture of constituents. It is the small, subtle differences in the mixture present in each area measured, together with the way those mixture differences reflect the illuminating light that results in the differences in the plotted data. This is true for any Earth surface cover, meaning that, it is a complex mixture of finer constituents that reflect electromagnetic energy in different ways. The details of that mixture are also characteristic of the material, in this case the soil type.

Thus what we are about is identifying mixtures to which we associate useful names, in this case the name of a specific soil type. What is needed is a way to characterize such mixture spectral responses as fully as possible.

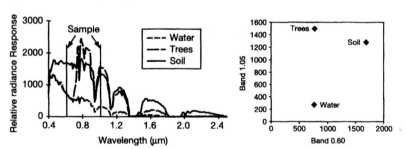

Figure 3-4. Data in spectral space and feature space.

Feature Space: From a continuous function to a discrete space

As has been seen, it is the job of the receptor in a pattern recognition system to measure the variables of the scene that are information bearing. According to our hypothesis, these are the variations present in the spectral response of Earth surface materials. We have identified the Receptor as a device that has a continuous spectral response function applied to its input and from it, produces a single vector-valued quantity at its output. Thus, the receptor transforms the information-bearing quantity, a spectral curve, from a continuous quantity to a discrete set of numbers or vector.

As seen in Chapter 1, a key factor in multispectral remote sensing is the matter of how one views these measurements. In particular, it was found there that it was helpful to view them in an N-dimensional space. That is, if the graphs on the left of Figure 3-4 represents the spectral responses of three classes of Earth cover and if the responses are sampled at only two wavelengths as indicated, then the two dimensional space on the right is the two dimensional space in which these two measurements are represented. That representation is a plot of the response at wavelength 2 (1.05 μm in this case) as a function of the response at wavelength 1 (0.60 μm here). The concept is that from the spectral presentation of the data in *Spectral Space*, the receptor has transformed significant features into *Feature Space*. Note that each continuous curve in the Spectral Space becomes a single point in Feature Space. Increasing the number of wavelengths sampled only increases the dimensionality of the Feature Space.

Then that family of curves making up a given class of material becomes a family of points or a distribution of points in Feature Space. And finally, note that as was seen in Chapter 1, not only the location but also the shape of this distribution is significant with regard to identifying the material. Much of the value of viewing data in Feature Space derives from this point. It is difficult to sense this aspect of a material's spectral response in Spectral Space, but in Feature Space the shape of this distribution and its relationship to any other materials sensed are very significant to the information extraction process.

In practice, the situation is complicated further by random variations induced by the atmosphere, noise in the sensor system itself, and other sources. Thus the actual signals to be dealt with will contain a combination of random-appearing variations that are diagnostic of the ground cover types to be identified and other random-appearing variations that are not. It will be necessary to carefully select techniques that are powerful enough to optimally discriminate between these "signals" despite these "noises."

What, then, if the two measurements such as in Figure 3-4 are not enough to distinguish among materials of interest in the scene? One could utilize measurements at more than two wavelengths in the same fashion. Though one cannot show more than three measurements graphically, requiring a three-dimensional depiction, they could certainly be dealt with computationally. Thus working in four or more dimensional spaces can be done in just the same manner, even though one can no longer draw a diagram showing how the data distribute. Indeed, if all wavelengths measured in Spectral Space are used, then the Feature Space contains all of the information present in the Spectral Space.

Summarizing to this point, it is assumed to be possible to identify the contents of each pixel of a multispectral image data set. The multispectral sensor serves as a pattern recognition receptor, providing a single point in a vector space called the Feature Space, for each pixel. The location and shape of the distribution of such points in the Feature Space are the characteristics that permit discriminating between those of the various classes of materials in the scene. But to do so, an economical way must be found to label (i.e., identify) the points in the Feature Space.

The Classifier: Partitioning the Space

To again illustrate the classifier concept, Figure 3-5 shows two simple algorithms for classifying a point in feature space. The basic idea is to divide the feature space into an exhaustive set of nonoverlapping regions, one for each class of interest, so that every point in the space is uniquely associated with one of the named classes. This implies establishing "decision boundaries." One begins with a set of pre-labeled points for each class. These are referred to as design samples or training samples. In the Minimum Distance classifier case, the identification of the unknown is done by determining the locus of points equidistant from the class means, thus defining a linear "decision boundary." In the Nearest-Neighbor case, it is done by determining the locus of points equidistant from the nearest members of two classes.

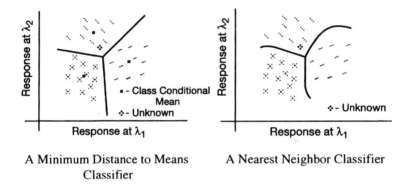

A Minimum Distance to Means
Classifier

A Nearest Neighbor Classifier

Figure 3-5. Two simple example classifier schemes.

Classification may not be so easy to do as in the case of these (oversimplified) diagrams, because (1) usually the classes desired have distributions in feature space that are not obviously separated, and (2) it is nearly always necessary to use more than three features. Thus one cannot "see" what the class distributions look like. The rules of conventional three-dimensional geometry do not apply in higher-dimensional space, as will be detailed in a later chapter. One

99

must have some kind of calculation procedure to determine the location of the decision boundaries required in order to define the needed mapping from feature space to the list of desired classes. The most common way to do this is by using discriminant functions.

3.3 Discriminant Functions

The decision boundaries in feature space, which divide the space into regions that are to be associated with each class, cannot be easily explicitly and directly definded. It involves a difficult computational problem of determining on which side of a boundary a given unknown pixel falls. There is an easier way. Assume that one can determine a set of m functions $\{g_1(x), g_2(x), \ldots, g_m(x)\}$

such that $g_i(x)$ is larger than all others whenever x is from class i.

This suggests that the appropriate classification rule is

Let ω_i denote the ith class.

Decide $x \in \omega_i$ iff (i. e., x is in class ω_i if and only if) $g_i(x) \geq g_j(x)$ for all $j = 1, 2, \ldots$ m.

The collection of functions $\{g_1(x), g_2(x), \ldots, g_m(x)\}$ are called discriminant functions. As shown in Figure 3-6, to classify an unknown pixel value x, one must only evaluate each discriminant

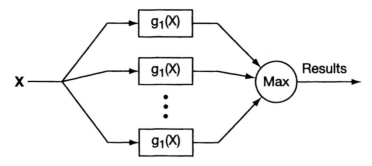

Figure 3-6. A pattern classifier defined in terms of discriminant functions, $g_i(x)$.

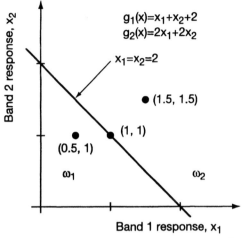

Figure 3-7. A Discriminant function example

function at **x** to determine which discriminant function is the largest to assign the pixel to a class, thus making the classification process simple and fast.

Figure 3-7 shows a simple example for a two-class case. In this case the two discriminant functions are $g_1(x)$ and $g_2(x)$ in the figure. The decision boundary in this case would be the line $x_1 + x_2 = 2$, that is, where $g_1(\mathbf{x}) = g_2(\mathbf{x})$, but using the discriminant function concept to classify samples, this never needs to be determined. Rather, evaluating these two functions at any point in the shaded area, such as, $\mathbf{x} = [0.5,1]^T$, results in g_1 being larger than g_2. Thus the decision rule would say decide in favor of class 1 for a pixel falling anywhere in this area. Similarly for points above the decision boundary, e.g., the point **x** $= [1.5, 1.5]^T$ where g_2 would be larger than g_1. The next question is, How does one define the discriminant functions, and how does one train the classifier?

3.4 Training the Classifier: An Iterative Approach

One method for determining the discriminant functions is to use the training samples iteratively. A procedure for doing so might be as follows.

    1. Choose a parametric form for the discriminant function, for example,

$$g_1(\mathbf{x}) = a_{11}x_1 + a_{12}x_2 + b_1$$
$$g_2(\mathbf{x}) = a_{21}x_2 + a_{22}x_2 + b_2$$

101

2. Initially set the a's and b's arbitrarily, for example, to $+1$ and -1

3. Sequence through the training samples, calculating the g's and noting the implied decision for each. When it is correct, do nothing, but when it is incorrect, augment (reward?) the a's and b's of the correct class discriminant, and diminish (punish?) that of the incorrect class. For example, if

$\mathbf{X} \in \omega_1$, but $g_2 > g_1$ then let

$$a'_{11} = a_{11} + \alpha x_1 \qquad\qquad a'_{21} = a_{21} - \alpha x_2$$
$$a'_{12} = a_{12} + \alpha x_2 \qquad\qquad a'_{22} = a_{22} - \alpha x_2$$
$$b'_1 = b_1 + \alpha \qquad\qquad b'_2 = b_2 - \alpha$$

4. Continue iterating through the training samples until the number of incorrect classifications is zero or adequately small.

Iterative training is basically the procedure used in, for example, training neural networks, only instead of a simple linear parametric model as above, a more complex configuration may be used. The potential advantage of doing so is its generality. The disadvantage in such neural net cases is that (1) it is entirely heuristic, and thus analytical evaluation and performance prediction is difficult and (2) generally, schemes in pattern recognition with large numbers of parameters require much larger training sets to properly estimate all the decision boundaries. While large training sets are possible in many pattern recognition applications, this is usually not the case in remote sensing. It is also the case that an iterative training procedure generally requires much more computation in the training phase than does a non-iterative parametric discriminant function approach. Thus such iterative training schemes are best applied in circumstances where re-training will be infrequent, something also not characteristic of remote sensing.

3.5 Training the Classifier: The Statistical Approach

An approach that avoids a number of the difficulties of iterative training methods is to use the laws of probability theory[10] and statistical estimation theory to determine suitable discriminant functions. A simple example will be used to introduce this approach.

Suppose one is to play a simple game with the following characteristics.[11]

Two players,

Two pairs of dice,

One normal pair

One Augmented pair (2 extra spots on each side)

Player 1 selects a pair of dice at random and rolls, announcing only the total showing.

Player 2 names which type of the pair of dice used, with a $1 bet.

To determine a "decision boundary" in this case, one can determine a list of all possible outcomes and how likely each is. Possible outcomes for a normal pair are the values 2 through 12; for an augmented pair, they are 6 through 16. Next consider how likely each value is. To obtain a 2 as the outcome, there is only one way, obtain a one on each die using a normal pair. For a three there are two ways, a 1 and a 2 or a 2 and a 1. For a four there are three ways,

---

10   Appendix A contains a brief outline of the Probability Theory, including most of the probability relations that will be needed in the remainder of this chapter. It is not essential that the reader is thoroughly familiar with this material, but it may help to briefly scan it if you are not otherwise acquainted with it.

11   This means for introducing concepts of statistical pattern recognition was originally devised by Prof. P. H. Swain in Chapter 3 of P.H. Swain and S.M. Davis, editors, *Remote Sensing, The Quantitative Approach*, McGraw-Hill, 1978.

Sum of Faces	Normal Dice			Augmented Dice		
	Possible Outcomes	# of outcomes	Probability	Possible Outcomes	# of outcomes	Probability
2	1,1	1	1/36	-	-	0
3	1,2 2,1	2	2/36	-	-	0
4	1,3; 2,2; 3,1	3	3/36	-	-	0
5	1,4; 2,3; 3,2; 4,1	4	4/36	-	-	0
6	1,5; 2,4; 3,3; 4,2; 5,1	5	5/36	3,3	1	1/36
7	1,6; 2,5; 3,4; 4,3; 5,2; 6,1	6	6/36	3,4; 4,3	2	2/36
8	2,6; 3,5; 4,4; 5,3; 6,2	5	5/36	3,5; 4,4; 5,3	3	3/36
9	3,6; 4,5; 5,4;6,3	4	4/36	3,6; 4,5; 5,4; 6,3	4	4/36
10	4,6; 5,5; 6,4	3	3/36	3,7; 4,6; 5,5; 6,4; 7,3	5	5/36
11	5,6; 6,5	2	2/36	3,8; 4,7; 5,6; 6,5; 7,4; 8,3	6	6/36
12	6,6	1	1/36	4,8; 5,7; 6,6; 7,5; 8,3	5	5/36
13	-	-	0	5,8; 6,7; 7,6; 8,5	4	4/36
14	-	-	0	6,8; 7,7; 8,6	3	3/36
15	-	-	0	7,8; 8,7	2	2/36
16	-	-	0	8,8	1	1/36
		36	1.0		36	1.0

Table 3-1. Table of ourcomes and probabilities.

a 1 and a 3, a 2 and a 2, or a 3 and a 1. Table 3-1 diagrams the possibilities.

Continuing this analysis, one finds that there are a total of 36 possible outcomes for the sums 2 through 12. One may thus calculate the probability of a two as $1/36 = 0.028$, a three as $2/36 = 0.056$, and so on. The result of this can be displayed as a histogram as in Figure 3-8. The result for the augmented set would be similar but shifted four units. The terminology used in the graph for p(x | standard) is read, "probability of the value x, given that the standard dice were used.

This pair of histograms, in effect, become the discriminant functions, and the decision boundary can then be set based on choosing the most probable outcome in any given case. For example, if a 7 is the outcome, though it could have come from either type, it is seen to have more likely come from the standard dice (6/36 to 2/36). If after rolling an unknown pair, a 4 shows, one can be assured that

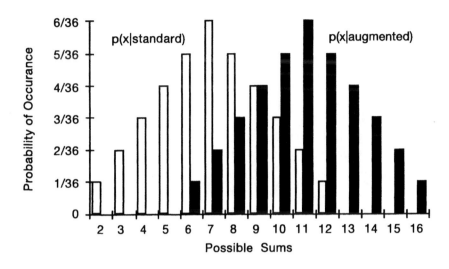

Figure 3-8. Histogram of the possible outcomes for the funny dice problem.

the dice are the standard pair since the probability of a 4 with the augmented pair is zero.

In this case the triangularly shaped probability distributions of the two possibilities were used as discriminant functions. These were arrived at by the assumption that the dice were "fair dice," rather than by any empirical determination. Thus training samples were not used. What if one could not assume that the dice were fair? Then one could roll the dice a number of times and keep track of how many times each possible outcome occurs. If one rolls the dice enough times and histograms the result, a histogram approximating the shape above should result. This is the situation that would need to be followed in a practical situation. We will follow this idea a bit further.

Suppose now that the data source was a single channel of a multispectral scanner. Instead of guessing which of two sets of dice was rolled, the objective might be to "guess" which of a set of ground-cover classes was being observed Again, the associated probability functions would be useful, and these could be estimated from training patterns. Given a set of measurements for a particular class, a tabulation would be made of the frequency with which each data value occurred for that class. The results could be displayed in the form of a histogram as shown in Figure 3-9. Similar histograms would be produced to estimate the probability functions for each class, and the histograms would then be used as in the "funny dice" example to assist in classifying data of unknown identity based on spectral measurement values.

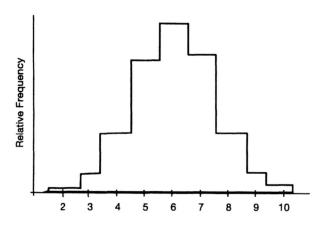

Figure 3-9. A Hypothetical histogram.

If the number of possible data values is large, storage in the computer of the histogram representation of the class probability functions may require a considerable amount of memory. Furthermore, if we try to generalize this approach to handle additional wavelength bands, the memory requirements will rapidly get out of hand; the number of memory locations needed to save an n-dimensional histogram in which each dimension can take on p values is p^n. One way to alleviate this problem is to assume that each histogram or probability function can be adequately approximated by a smooth curve having a simple functional form. For example, assume that the probability function for a class of interest can be approximated by a normal (i.e., Gaussian) probability density function, an assumption for which we will later provide justification. For the one-dimensional or univariate case, the normal density function for class i is given by

$$p(x|\omega_i) = \frac{1}{\sqrt{2\pi\sigma_i^2}} \exp\left[\frac{-(x-\mu_i)^2}{2\sigma_i^2}\right]$$

where exp [ ] = e (the base of the natural logarithms) raised to the indicated power

$\mu_i = E[x \mid \omega_i]$ is the mean or average value of the measurements in class i

$\sigma_i^2 = E[(x - \mu_i)^2 \mid \omega_i]$ is the variance of the measurements in class i

The key characteristics of the Gaussian density function are that the mean value occurs at the peak of the density function, that it is

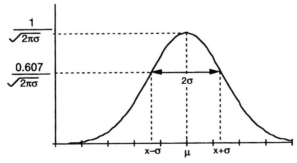

Figure 3-10. Univariate Gaussian probability density function.

107

symmetrical about its mean, and that the width of the density function is proportional to σ_i, the variance. This is illustrated in Figure 3-10.

It is also the case that the area under any probability density function must be one. Thus, as σ increases and the width becomes larger, the height of the central value becomes smaller. By choosing the value of the mean, μ, to be centered on the histogram, and choosing the width of the density function by setting the variance, σ, to be that determined from the histogram, one can reasonably approximate the shape of the histogram with a Gaussian density. Figure 3-11 shows the result of doing so graphically.

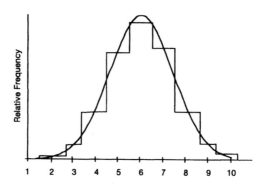

Figure 3-11. Fitting a Gaussian curve to a histogram.

Usually for spectral data there would be many more than 10 possible values, so the steps in the histogram will be smaller and more frequent, and, μ_i and σ_i^2 will be unknown and must be estimated from training samples. Thus, in principle, we now have a very practical way to design a pattern classifier for multispectral data analysis. One must decide on a list of classes, and then select training samples for each class to quantify the selections. Using the training samples, estimate the mean vector and covariance matrix for each class. The density functions defined by this set of mean vectors and covariance matrices then constitute the discriminant functions for the classifier.

But, is it reasonable to use a Gaussian density function to approximate the distribution of measurements for every class? The answer, of course, is no. However, what can be done is to approximate an arbitrarily shaped class distribution by a linear combination of appropriately chosen Gaussian densities. For example, the non-Gaussian density (solid curve) shown in Figure 3-12 may be represented by the sum of two Gaussian densities (dotted curves) using appropriate weighting between the two. The great simplicity that the Gaussian density provides, particularly in higher-dimensional cases, continues to make it attractive. Thus this Gaussian mixture approach for modeling complex classes continues to be appropriate for our purposes. We shall return to the matter of dealing with non-Gaussian class densities later.

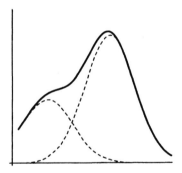

Figure 3-12. A non-Gaussian density (solid curve) represented
by the sum of two Gaussian densities (dotted curves).

We next want to add some details to the matter of classifier design. Let us complicate the funny dice rules. Suppose that the dice to be rolled are selected at random from 80 standard pairs and 20 augmented pairs. You bet $1 on each play:

- If you guess wrong, you lose your dollar.
- If you guess correctly when a <u>standard</u> pair is drawn, you win your dollar.
- If you guess correctly when an <u>augmented</u> pair is drawn, you win $5.

To determine if you want to play, let us seek to determine the winnings or losses after 100 plays.

There are several possible strategies to use in devising a way for the pattern recognition algorithm to make the desired classifications under these circumstances. Possible strategies are as follows:

Maximum Likelihood strategy[12]

$g_s^1(\mathbf{x}) = p(\mathbf{x} \mid S)$ i.e., the probability of the value $\mathbf{x}$, given class S,

$g_a^1(\mathbf{x}) = p(\mathbf{x} \mid A)$

That is, use the probability histograms directly as the discriminant functions. This yields the most likely class and is referred to as a maximum likelihood classifier.

3.5.2. Minimum A posteriori Probability of Error Strategy

If it is the probability of a correct choice that one wishes to maximize, one should base the decision on maximizing $p(\omega_i \mid \mathbf{x})$, the so-called *a posteriori* probability, the probability of class i after knowing the value of $\mathbf{x}$. Bayes's Theorem states that since $p(\mathbf{x}, \omega_i) = p(\omega_i \mid \mathbf{x})p(\mathbf{x}) = p(\mathbf{x} \mid \omega_i)p(\omega_i)$, meaning,, the probability of $\mathbf{x}$ and ω_i jointly occurring is equal to the probability that ω_i occurs,

---

[12] We will be describing three different strategies for classification here, this being the first of the three. Thus the superscript 1 on g_s here designates this as the first of the three.

given that the value **x** has occurred, or the probability that the value **x** occurs, given that it is known ω_i occurred, each multiplied by the probability of the known event. Then what is desired is to maximize the probability that ω_i occurred given that the value **x** was measured, or

$$p(\omega_i|\mathbf{x}) = \frac{p(\mathbf{x}|\omega_i)p(\omega_i)}{p(\mathbf{x})} = \frac{p(\mathbf{x},\omega_i)}{p(\mathbf{x})}$$

Note that the denominator term, $p(\mathbf{x})$, is the same for any i, thus, for the problem at hand, pick the larger of $g_i{}^2(\mathbf{x})$, where,

$$g_s{}^2(x) = p(x,S) = p(x|S)p(S) = p(x|S)*0.8$$
$$g_a{}^2(x) = p(x,A) = p(x|A)p(A) = p(x|A)*0.2$$

This strategy is referred to as a Minimum A Posteriori Error Strategy (MAP) or the <u>Bayes's</u> <u>Rule</u> <u>Strategy</u>. Note that for equal class prior probabilities $p(\omega_i)$, this reduces to the Maximum likelihood strategy.

Minimum Risk or Minimum Loss Strategy

If different results have different consequences and one can quantify the consequence (loss) in each case, one could seek to minimize the expected loss. The <u>expected</u> <u>loss</u> for each of the L classes is given by

$$L_{\underline{\omega}} = \sum_{\omega_i=1}^{L} \lambda(\underline{\omega}\,|\,\omega_i)p(\omega_i\,|\,\mathbf{x})$$

where λ is the loss on any one outcome. ($\underline{\omega}$ is your answer, ω_i the true result.) That is, it is the sum of the losses for each outcome times the probability that that outcome occurs. Thus in the case at hand,

$$L_{\underline{S}}(x) = \lambda(\underline{S} \mid S)\, p(S \mid x) + \lambda(\underline{S} \mid A)\, p(A \mid x)$$
$$L_{\underline{A}}(x) = \lambda(\underline{A} \mid S)\, p(S \mid x) + \lambda(\underline{A} \mid A)\, p(A \mid x)$$

For the discriminant functions, one can use

$$g_s^3(x) = -L_{\underline{S}}(x)$$
$$g_A^3(x) = -L_{\underline{A}}(x)$$

For this problem, $\lambda(\underline{S}|S) = -1$ (a win = a negative loss), $\lambda(\underline{S}|A) = 1$, $\lambda(\underline{A}|S) = 1$, and $\lambda(\underline{A}|A) = -5$

Or one can use the equivalent discriminant functions:

$$g_{\omega}^3(x) = -\sum \lambda(\underline{\omega} \mid \omega_i)\, p(x \mid \omega_i)\, p(\omega_i)$$

which yields

$$g_S^3(x) = \ p(x \mid S)\, p(S) - p(x \mid A)\, p(A)$$
$$g_A^3(x) = -\, p(x \mid S)\, p(S) + 5\, p(x \mid A)\, p(A)$$

The next step is to determine the decision rules. To do so it is convenient to establish a spreadsheet table, as in Table 3-2.

Table 3-2. The Decision Table

	Max Likelihood			Max. a posteriori			Minimum Risk		
x	p(x|S)	p(x|A)	Decision	p(x|S)p(S)	p(x|A)p(A)	Decision	$g_s3(x)$	$g_A3(x)$	Decision
1	0	0		0	0				
2	0.028	0	S	0.022	0	S	0.022	-0.022	S
3	0.056	0	S	0.044	0	S	0.044	-0.044	S
4	0.083	0	S	0.067	0	S	0.067	-0.067	S
5	0.111	0	S	0.089	0	S	0.089	-0.089	S
6	0.139	0.028	S	0.111	0.006	S	0.106	-0.083	S
7	0.167	0.056	S	0.133	0.011	S	0.122	-0.078	S
8	0.139	0.083	S	0.111	0.017	S	0.094	-0.028	S
9	0.111	0.111	S	0.089	0.022	S	0.067	0.022	S
10	0.083	0.139	A	0.067	0.028	S	0.039	0.072	A
11	0.056	0.167	A	0.044	0.033	S	0.011	0.122	A
12	0.028	0.139	A	0.022	0.028	A	-0.006	0.117	A
13	0	0.111	A	0	0.022	A	-0.022	0.111	A
14	0	0.083	A	0	0.017	A	-0.017	0.083	A
15	0	0.056	A	0	0.011	A	-0.011	0.056	A
16	0	0.028	A	0	0.006	A	-0.006	0.028	A

From the table the following decision rules can be established.

- Maximum Likelihood: Decide
  $d_1(x)$ $X \in S$ for $2 \leq x \leq 9$,
  $X \in A$ for $10 \leq x \leq 16$

- Minimum Error Probability: Decide
  $d_2(x)$ $X \in S$ for $2 \leq x \leq 11$,
  $X \in A$ for $12 \leq x \leq 16$

  - Minimum Risk: Decide
  $d_3(x)$ $X \in S$ for $2 \leq x \leq 9$,
  $X \in A$ for $10 \leq x \leq 16$

(It is coincidental that Maximum Likelihood and Minimum Risk are the same.)

To evaluate the three strategies, calculate the probability of being correct on any given play and the expected winnings after 100 plays. This may be done as follows.

$$\text{pr(correct)} = \sum^{\omega_i} \text{pr}(\mathbf{x} \in R_i \,|\, \omega_i) \quad R_i : g_i(\mathbf{x}) \geq g_j(\mathbf{x}) \; i \neq j$$

For the maximum likelihood and minimum risk cases:
$$\text{pr(correct)} = \text{pr}(x \leq 9 \text{ and } \omega = S) + \text{pr}(x \leq 10 \text{ and } \omega = A)$$

$$= \sum_{x=2}^{9} \text{pr}(x, S) + \sum_{x=10}^{16} \text{pr}(x, A)$$

$$= \sum_{x=2}^{9} \text{pr}(x \,|\, S)\text{pr}(S) + \sum_{x=10}^{16} \text{pr}(x \,|\, A)\text{pr}(A)$$

A similar equation is used for the minimum error rule except summing over the limits defined by that decision rule.

For the winnings after 100 plays, calculate the expected (average) loss per play using the previous loss formula, then multiply by -1 (to get expected winnings per play instead of loss) and 100. This can be done as follows: For a given class ω_i and decision rule, $d(x)$, the expected loss is

$$L_{d(x)}(\mathbf{x}) = \sum^{\omega_i} \lambda(d(\mathbf{x})\,|\,\omega_i)p(\omega_i\,|\,\mathbf{x})$$

Then the expected loss over all possible outcomes (i.e., classes and x's) is

$$E\big[L_{d(x)}(\mathbf{x})\big] = \sum_{x=2}^{16} L_{d(x)}p(\mathbf{x})$$

$$= \sum_{x=2}^{16}\sum^{\omega_i} \lambda(d(\mathbf{x})\,|\,\omega_i)p(\omega_i\,|\,\mathbf{x})p(\mathbf{x})$$

$$= \sum_{x=2}^{16}\sum^{\omega_i} \lambda(d(\mathbf{x})\,|\,\omega_i)p(\mathbf{x}\,|\,\omega_i)p(\omega_i)$$

The results for both p(correct) and the expected winnings turn out as follows:

Maximum Likelihood and Minimum Risk
p(correct) = 0.811 $/100 plays = +$120

Minimum Error Probability
p(correct) = 0.861 $/100 plays = + $105.55

Thus, although the latter decision rule will produce more correct responses, the former decision rules will produce a larger dollar win. However, remember that it is coincidental that the Maximum Likelihood strategy also produced minimum risk in this case.

3.6 Discriminant Functions: The Continuous Case

The funny dice example was a case involving discrete probabilities. That is, the possible outcomes were a small number of discrete values, the integers from 2 to 16 such that each possible outcome could be described in terms of a discrete-valued probability. The usual case for multispectral remote sensing is that the measurement of pixel response values can result in so many different discrete digital values that it is more convenient to consider the measurement values as being a continuum.

This way, instead of a histogram of probability to describe a class as was done above, a class probability density function would be used. The modeling of each class is simply a modeling of the probability density function for that class in feature space. This turns out to be a key step. How well a classifier will work depends very much on how well the true class density functions can be determined from the training samples.

3.7 The Gaussian Case

The Gaussian density function is perhaps the most widely studied density function for any purpose, because it has so many convenient properties and fits so many processes in nature. The Central Limit Theorem states in general terms that if a random observable is made up of the sum of a large number of independent random quantities, each with the same distribution, the observable will be Gaussian, regardless of the distribution governing the constituent random quantities. This approximates many circumstances that describe

115

how spectral responses are formed. Also, among the other properties that make the Gaussian density useful as a class model is the fact that it is parametric in form with a small number of parameters that are easy to relate to real world observables.

As previously noted, the univariate Gaussian density function is given by

$$p(x \mid \omega_i) = \frac{1}{\sqrt{2\pi}\sigma_i} \exp\left[\frac{-(x-\mu_i)^2}{2\sigma_i^2}\right]$$

One must estimate only μ_i and σ_i in order to have a description of the density function over its entire range. But what if there are two or more channels of multispectral data? One of the very convenient features of the Gaussian density function is that it generalizes so conveniently to the multidimensional case. In this case, assuming n dimensions, the density may be written in vector form (boldfaced symbols indicate vector quantities) as

$$p(\mathbf{x} \mid \omega_i) = (2\pi)^{-n/2}|\Sigma_i|^{-1/2} \exp\left\{-\frac{1}{2}(\mathbf{x}-\mu_i)^T \Sigma_i^{-1}(\mathbf{x}-\mu_i)\right\}$$

where

$$\mathbf{x} = \begin{bmatrix} x1 \\ x2 \\ \cdot \\ \cdot \\ \cdot \\ xn \end{bmatrix} \quad \mu_i = \begin{bmatrix} \mu1 \\ \mu2 \\ \cdot \\ \cdot \\ \cdot \\ \mu n \end{bmatrix} \quad \Sigma_i = \begin{bmatrix} \sigma_i11 & \sigma_i12 & \cdots & \sigma_i1n \\ \sigma_i21 & \sigma_i22 & \cdots & \sigma_i2n \\ \cdot & \cdot & \cdots & \cdot \\ \cdot & \cdot & \cdots & \cdot \\ \cdot & \cdot & \cdots & \cdot \\ \sigma_in1 & \sigma_in2 & \cdots & \sigma_inn \end{bmatrix}$$

the vector valued measurement value, the mean vector, and the covariance matrix, respectively. Then, in the expression for $p(x|\omega_i)$, $|\Sigma_i|$ is the determinant of the covariance matrix, Σ_i^{-1} is the inverse of Σ_i, and $(x - \mu_i)^T$ is the transpose of $(x - \mu_i)$. For notational convenience, the Gaussian density function is often written as

$$p(x|\omega_i) \sim N(\mu_i, \Sigma_i)$$

The mean vector, μ_i, is called a first-order statistic, because it is a function of a random quantity in one band at a time. The covariance matrix, σ_i, is referred to as a second-order statistic, because each element of it is a relationship between pairs of bands.

An additional computation convenience is the following. For the Minimum A Posteriori error case, the decision rule becomes

Decide $X \in \omega_i$ iff
$$p(x|\omega_i)\,p(\omega_i) \geq p(x|\omega_j)\,p(\omega_j) \text{ for all } j = 1,2,...,m$$

Now if $p(x|\omega_i)p(\omega_i) \geq p(x|\omega_j)p(\omega_j)$ for all $j = 1,2,...,m$, then it is also true that

$$\ln p(x \mid \omega_i)p(\omega_i) \geq \ln p(x \mid \omega_j)p(\omega_j) \text{ for all } j = 1,2,...,m$$

Thus we may take the following as an equivalent discriminant function, but one that requires substantially less computation time. (Note in this expression that the factor involving 2π has been dropped, as has a leading factor of 2 in the final form below, since they would be common to all class discriminant functions and thus would not contribute to the discrimination.)

$$g_i(x) = \ln p(\omega_i) - \frac{1}{2}\ln|\Sigma_i| - \left\{ \frac{1}{2}(x - \mu_i)^T \Sigma_i^{-1}(x - \mu_i) \right\}$$

or
$$g_i(x) = \ln \frac{p^2(\omega_i)}{|\Sigma_i|} - (x - \mu_i)^T \Sigma_i^{-1}(x - \mu_i)$$

Note also that the first term on the right must only be computed once per class, and only the last term must be computed for each measurement to be classified.

Unbiased multidimensional (vector) estimators for the Gaussian density parameters are

$$\hat{\mu}_i = \frac{1}{q_i} \sum_{k=1}^{q_i} x_k$$

$$\sigma_{ijk} = \frac{1}{q_i - 1} \sum_{l=1}^{q_i} (x_{jl} - \hat{\mu}_j)(x_{kl} - \hat{\mu}_k) \quad j,k = 1,2,...n$$

In the two dimensional case, the density takes on the three-dimensional bell shape. In more than two dimensions, we can no longer draw a diagram of its shape. However, even in the multidimensional case, the Gaussian density continues to have the key characteristics, namely the mean value occurs at the peak of the density function, it is symmetrical about its mean, and the width of the density function in each dimension is proportional to σ_i, the variance in that dimension. The off-diagonal terms of the covariance matrix describe the eccentricity of the density. Basically the mean vector then defines the location of the density in n-dimensional space, while the covariance matrix provides information about its shape.

3.8 Other Types of Classifiers
There is now a large array of different types of classifier algorithms that appear in the literature. A hierarchically arranged list of examples follows. It is assumed here that there are m classes, and, for simplicity of present purposes, the classes are assumed equally likely.

Ad hoc and Deterministic Algorithms

The nature of variations in spectral response that are usable for discrimination purposes is quite varied. Diagnostic spectral characteristics may extend all the way from the general shape of the entire response function in spectral space, spread across many bands, to very localized variations in one or a small number of narrow spectral intervals. Many algorithms have appeared in the literature that are designed to take advantage of specific characteristics on an ad hoc basis. Example algorithms of this type extend from simple parallelepiped algorithms to spectral matching schemes based on least squares difference in spectral space between an unknown pixel response that has been adjusted to reflectance and a known spectral response from a field spectral database, on to an imaging spectroscopy scheme based on one or more known molecular absorption features.

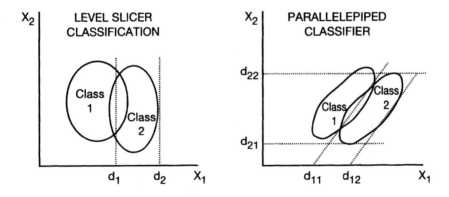

Figure 3-13. Two types of simple ad hoc classifiers.

Among the wide assortment of classifier algorithms that have found use in remote sensing, perhaps the simplest is the Level Slicer, illustrated in Figure 3-13. It has the advantage of simplicity, but its usefulness is only in extremely limited, simple cases. The next step up in complexity would be the Parallelepiped classifier, which is essentially a level slicer where the slice level is made conditional upon the value of a second feature.

Such algorithms are sometimes motivated by a desire to take advantage of perceivable cause-effect relationships. These algorithms are usually of a nature that the class is defined by a single spectral curve,namely a single point in feature space. Where this is the case, they cannot benefit from or utilize second-order class information.

Minimum Distance to Means

$$g_i(\mathbf{x}) = (\mathbf{x} - \mu_i)^T (\mathbf{x} - \mu_i)$$

choose class i iff $\quad g_i(\mathbf{X}) \leq g_j(\mathbf{X})$ for all $j = 1, 2, \ldots m$

In this case, pixels are assigned to whichever class has the smallest Euclidean distance to its mean (note that the inequality symbol in the decision rule is reversed in this case). The classes are, by default, assumed to have probability density functions with common covariances that are equal to the identity matrix. This is equivalent to assuming that the classes all have unit variance in all features and the features are all uncorrelated to one another.

Geometrically, for the two-class, two-feature case, Figure 3-14 shows how the decision boundaries for this classifier would appear for a given set of classes distributed as the two elongated oval areas. It is seen in this case that the decision boundary for the Minimum Euclidean Distance classifier is linear and is in fact the perpendicular bisector of the line between the two class mean values. Its location is uninfluenced by the shapes of the two class distributions. The

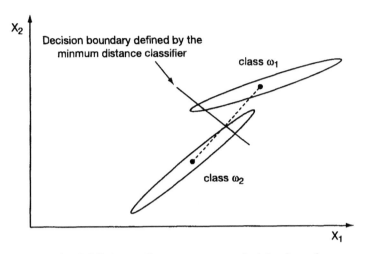

Figure 3-14. Minimum distance to means decision boundary
for an example two class, two feature case.

gray areas indicate the pixels that would be classified in error in
this case. Note that if there were a classification scheme that
accounted for the shape of the classes here in addition to the location
of the classes, in terms of their centroid (mean value), a more
accurate result could be obtained.

Fisher's Linear Discriminant

$$g_i(\mathbf{x}) = (\mathbf{x} - \mu_i)^T \Sigma^{-1} (\mathbf{x} - \mu_i)$$

choose class ω_i iff $\quad g_i(\mathbf{x}) \le g_j(\mathbf{x})$ for all $j = 1, 2, \ldots m$.

In this case, the classes are assumed to have density functions with
a common covariance specified by Σ. This is equivalent to assuming
that the classes do not have the same variance in all features, the
features are not necessarily uncorrelated, but all classes have the
same variance and correlation structure. In this case the decision
boundary in feature space will be linear, but its location and

121

orientation between the class mean values will depend on Σ, the combined covariance for all the classes in addition to the class means.

Geometrically, for the two-class, two-feature case, Figure 3-15 shows how the decision boundary for this classifier would appear for a given set of classes indicated by the two elongated oval areas. It is seen in this case that the decision boundary for the Fisher Linear Discriminant classifier is also linear but its orientation and location are influenced by the overall shape of the combined distribution. The gray areas again indicate pixels that would be classified in error.

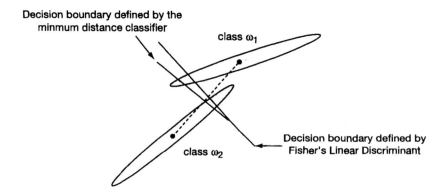

Figure 3-15. Fisher Linear Discriminant decision boundary for an example two-class, two-feature case.

Quadratic (Gaussian) Classifier

$$g_i(\mathbf{x}) = - (1/2)\ln|\Sigma_i| - (1/2)(\mathbf{x}-\mu_i)^T\Sigma_i^{-1}(\mathbf{x}-\mu_i)$$

choose class ω_i iff $\quad g_i(\mathbf{x}) \geq g_j(\mathbf{x})$ for all $j = 1,2,\ldots m$

122

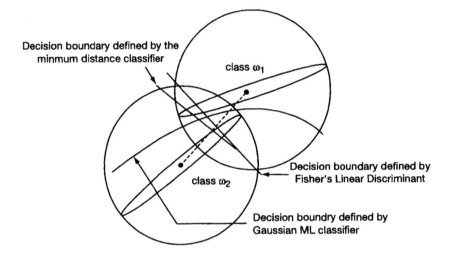

Figure 3-16. Decision boundaries for several classifier algorithms
for an example two-class, two-feature case.

In this case the classes are not assumed to have the same covariance, each being specified by Σ_i. The decision boundary in feature space will be a second-order hypersurface (or several segments of second order hypersurfaces if more than one subclass per class is assumed), and its form and location between the class mean values will depend on the Σ_i's.

Geometrically, for the two-class, two-feature case, Figure 3-16 shows how the decision boundaries for this classifier would appear for a given set of classes indicated by the two elongated oval areas.

Interestingly, if the data for the two classes had the mean values and class and feature variances as indicated, but with features uncorrelated with one another, the data would be spread over the regions indicated by the circles. Under these circumstances the error rate, as indicated by the overlapped area would be substantially greater. This is another illustration of the fact that correlation does not always imply redundancy and features being uncorrelated is not necessarily desirable.

Note that as described, the minimum distance to means, the Fisher linear discriminant, and the quadratic classifier are all maximum likelihood classifiers. They form a hierarchically related set and differ only in the assumptions about the details of the class covariance functions. Not apparent from this discussion is the influence of the number of training samples and therefore the precision of the estimates of the mean vectors and covariance matrices, as determined from the training samples has on the process. Though the value of increasing classifier complexity seems apparent, this must be balanced against the increasing difficulty with which to obtain precise estimates of the parameters describing the class distributions.[13]

3.8.5. Nonparametric Methods

Perhaps the next step up in classifier complexity would be a "nonparametric" classifier. A nonparametric scheme is one that does not assume a particular form for the class-conditional density functions. The *Parzen Window Classifier* is an example. In this case, one seeks to approximate the class density functions in terms of a series of window functions or kernels. The discriminant function in this case is

$$g_i(\mathbf{x}) = \frac{1}{N_i} \sum_{j=1}^{N_i} K(\frac{\mathbf{x} - \mathbf{x}_{ij}}{\lambda})$$

---

[13] Raudys, S.J. and A.K. Jain, Small Sample Size Effects in Statistical Pattern Recognition: Recommendations for Practitioners, *IEEE Transactions on Pattern Analysis and Machine Intelligence*, Vol. 13, No. 3, pp. 252-264, March 1991.

Choose class ω_l iff $\quad g_i(\mathbf{x}) \geq g_j(\mathbf{x})$ for all $j = 1, 2, \ldots m$

Here $K(\cdot)$ is the kernel (window function), λ is a smoothing parameter and $\mathbf{x}_{ij}$ is a location parameter for a given class and term of the summation. A popular window function is the exponential window

$$K_e\left(\frac{\mathbf{x} - \mathbf{x}_{ij}}{\lambda}\right) = \exp\left(-\frac{\left(\mathbf{x} - \mathbf{x}_{ij}\right)^T\left(\mathbf{x} - \mathbf{x}_{ij}\right)}{\lambda^2}\right)$$

It is seen that this window function is the Gaussian kernel. In this case the (nonparametric) class-conditional density function is being modeled by a linear combination of Gaussian functions. From this viewpoint, one can see that even if, in a practical circumstance, a class at hand is clearly not Gaussian, it may be satisfactorily represented in terms of a linear combination of Gaussian "subclasses." This turns out to be a very powerful, practical approach, since it can have all the advantages of a parametric scheme and yet is adequately general. This then may be referred to as a "semiparametric" scheme.

Another scheme that does not assume a parametric form for the class-conditional densities is the K-Nearest Neighbor classifier. In this case the unknown vector is assigned to the class of the majority of its K nearest neighbors. The K-NN rule is somewhat similar to a Parzen window classifier with a hyper-rectangular window function.

Nonparametric classifiers take on many forms, and their key attractive feature is their generality. As represented above, $K(\cdot)$ is a kernel function that can take on many forms. The entire discriminant function has N_i terms, each of which may contain one or more arbitrarily selected parameters. Thus, the characteristic that gives a nonparametric scheme its generality is this often large number of

125

parameters. However, every detailed aspect of the class density must be determined by this process, and this can quickly get out of hand.

At first look, then, it might seem obvious that the approach that uses the greatest amount of information in this sense from the design set would be the best. However, as will be seen shortly, a more thorough understanding of the situation makes this assumption incorrect, because, especially in remote sensing, design sets are almost always of limited size. We will explore this aspect in depth shortly.

Other Special Purpose Algorithms

There also have emerged a number of variants to this generic list of hierarchical schemes. Two additional schemes follow.

A classifier called the **Correlation Classifier** has discriminant function

$$g_i(\mathbf{x}) = \left(\frac{\mathbf{x}^T \mu_i}{\sqrt{\mathbf{x}^T \mathbf{x}} \sqrt{\mu_i^T \mu_i}} \right)$$

Note that this classifier again does not use covariance information. The denominator normalizes the quantity inside the brackets; therefore the discriminant value ranges between minus one and one. Since this discriminant function varies between minus one and one, a variation on the correlation classifier is the **Spectral Angle Mapper (SAM)**:

$$g_i(\mathbf{x}) = \cos^{-1} \left(\frac{\mathbf{x}^T \mu_i}{\sqrt{\mathbf{x}^T \mathbf{x}} \sqrt{\mu_i^T \mu_i}} \right)$$

The discriminant function thus may be seen as an angle.

A **Matched Filter** classifier called **Constrained Energy Minimization** has discriminant function as follows:

$$g_i(\mathbf{x}) = \frac{\mathbf{x}^T \mathbf{C}_b^{-1} \mu_i}{\mu_i^T \mathbf{C}_b^{-1} \mu_i}$$

where $\mathbf{C}_b$ is the common correlation matrix, or the correlation matrix of the background scene.

We again note that methods that use multiple samples for training have a substantial advantage over deterministic methods that utilize only a single spectrum to define a class or only the mean value of several spectra. The latter tend to require very high signal-to-noise ratios, whereas those based on multiple sample training sets tend to be more immune to the effects of noise.

The means for quantitatively describing a class distribution from a finite number of training samples commonly comes down to estimating the elements of the class mean vector and covariance matrix, as has been seen. Sound practice dictates that the number of training samples must be large compared to the number of elements in these matrices. When the number of training samples is limited, as it nearly always is in remote sensing, and the dimensionality of the data becomes large, the needed relationship between the training set size and the number of matrix elements that must be estimated quickly becomes strained even in the parametric case. This is especially true with regard to the covariance matrix, whose element population grows very rapidly with dimensionality. This point will be explored in more detail in a later chapter.

3.9 Thresholding

In a practical remote sensing circumstance, in addition to the main classes present in a scene, there may be a number of minor classes in which there may be no interest. For example, in classifying an agricultural area, one may not be interested in a class for houses, since they would occupy only a small proportion of scene area and thus few pixels. They might also be expected to have a substantially different spectral response than the main crops classes and thus occur well out on the tails of all class distributions. In order to conveniently deal with such minor classes, consider the following:

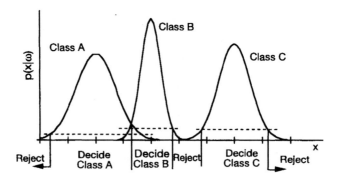

Figure 3-17. An illustration of the thresholding concept.

The probability density functions for three classes in one dimension are shown in Figure 3-17. To take care of minor classes which might be off the scale to the left or right, define a threshold level for each class as shown by the dotted lines. Redefine the decision rule to

decide $\mathbf{x} \in \omega_i$ iff $\qquad g_i(\mathbf{x}) \geq g_j(\mathbf{x})$ for all $j = 1, 2, \ldots m$

and $\qquad g_i(\mathbf{x}) > T_i$

Pixels that do not meet these criteria for any class would be assigned to a null class.

Note that for the Gaussian case, this would mean that

$$\ln p(\omega_i) - (1/2)\ln|\Sigma_i| - (1/2)(x - \mu_i)^T\Sigma_i^{-1}(x - \mu_i) > T_i$$

therefore

$$(x - \mu_i)^T\Sigma_i^{-1}(x - \mu_i) < -2T_i - 2\ln p(\omega_i) - \ln|\Sigma_i|$$

The left side of this expression is X^2 (i.e. Chi-squared) distributed with n degrees of freedom. Thus, for a true Gaussian class, it is possible to choose a percent of the measurement vectors to be thresholded and look up in a X^2 table what threshold value is needed to achieve it.[14] For example, if 5% of the measurements of a four feature class are to be rejected, the table gives a value of 9.488. Thus,

$$T_i = -4.744 + \ln p(\omega_i) - (1/2)\ln|\Sigma_i|$$

3.10 On The Characteristics, Value, and Validity of the Gaussian Assumption

When it is appropriate, the Gaussian assumption has a significant advantage over a nonparametric approach. However, it is often difficult to tell when the Gaussian assumption is appropriate. Although one can examine the histogram of the training samples of each class in each band, this can be misleading. This has to do, among other things, with the number of training samples available

---

[14] For further details on Chi-squared values, the reader may wish to consult Anderson, T. W., *An Introduction to Multivariate Statistical Analysis*, John Wiley and Sons, New York, 1958, p.112

for estimating the class distributions and the number of possible gray values (discrete bins) in each band.

Histogram of Test Scores

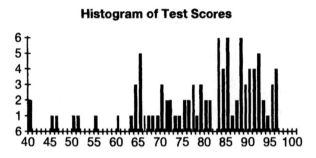

Figure 3-18. Histogram of the scores of 97 students.

Consider the example of Figure 3-18, taken from the grades scored on a test given to 97 students. Though class scores are often assumed to be "normally" distributed (i.e., Gaussian), these scores do not appear to be particularly Gaussian. But suppose that the histogram is re-plotted with a bin size of five instead of one, as shown in Figure 3-19. Now it appears the data may be made up of two and perhaps three Gaussian densities, one centered at 85, another at 65, and perhaps a third at 45. The problem initially was that for the level of detail being used,namely, the number of bins being used in the histogram, there were not enough data available to achieve a good estimate of the shape of the distribution. This is a first indication of an issue with measurement complexity that will be treated in more detail below.

For the moment however, as far as statistics estimation is concerned, it is usually the case that a reasonably accurate estimate of the mean and standard deviation (or in the multivariate case, the covariance matrix) can be made with far fewer samples than those needed to estimate the entire shape of the distribution. Thus, since it is a characteristic of pattern recognition methods applied to the remote

130

Histogram by Fives

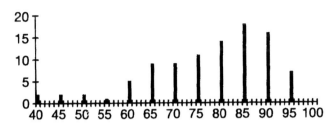

Figure 3-19. Histogram of the tests scores of Figure 3-18, but
with a bin size of 5.

sensing problem that there are never as many training samples available as might be desired, this can be very important. To illustrate, one may try training a funny dice classifier by rolling a pair of dice many times, plotting the histogram of the outcomes, and estimating the mean and standard deviation after each roll. One will likely find that it takes many more rolls of the dice to achieve a histogram as that of Figure 3-8, if one tries to design a classifier for that problem from empirically derived training data rather than from the theoretical assumptions of fair dice used above.

As to how much data are needed, note the following:
- Funny Dice – 11 bins
- One band, 6 bit data – $2^6 = 64$ bins
- Landsat MSS, 4 bands – $64^4 = 16,777,216$ bins
- Landsat TM, 7 bands, 8 bit – $256^7 = 7.2 \times 10^{16}$ bins

It is seen that as the dimensionality goes up, the potential number of bins for which likelihood must be estimated goes up very fast. This greatly increases the requirement for larger and larger numbers of training samples if good performance is to be obtained. This situation will be investigated more formally in the following section.

3.11 The Hughes Effect

The relationship between the number of discrete bins in feature space and the number of training samples has been formally derived on a theoretical basis.[15] One of the results of this study is given in Figure 3-20. The variable of the vertical axis of this figure is the mean recognition accuracy obtainable from a pattern classifier, averaged over all possible pattern classifiers. This is plotted as a function of measurement complexity on the horizontal axis. In this case, measurement complexity is a measure of how complex and detailed a measurement is taken. For digital multispectral data, it corresponds to the number of bins or brightness-level values recorded, k, raised to the nth power, where n is the number of spectral bands, i.e., the number of possible discrete locations in n-dimensional feature space. The more bands one uses and the more

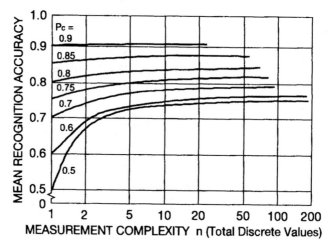

Figure 3-20. Mean Recognition Accuracy vs. Measurement

---

[15] G. F. Hughes, On the mean accuracy of statistical pattern recognizers, *IEEE Transactions on Information Theory,* Vol. IT-14, No. 1, January 1968.

brightness levels in each band, the greater the measurement complexity. The assumption was made in the referenced work that there are only two classes to choose between and the parameters used in the figure is P_C, the *a priori* probability of occurrence of one of the two classes.

The results show, as one might expect, that as one increases the measurement complexity, one can always expect the accuracy to increase. However, there is a saturating effect that occurs, in that after a certain measurement complexity is reached, the increase in accuracy with increasing measurement complexity is very small.

Figure 3-20 is the result one obtains if an infinite number of training samples are available by which to train the pattern classifier so as to be able to perfectly determine the class distributions governing each class. During the same study, Dr. Hughes obtained the result for a finite number of training samples; this is shown in Figure 3-21. In this figure, the two classes are assumed equally likely and

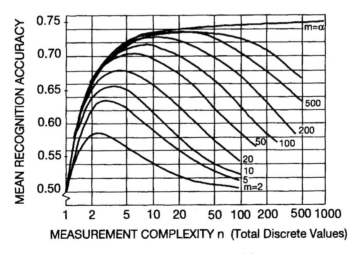

Figure 3-21. Mean Recognition Accuracy vs. Measurement
Complexity for the finite training case.

133

the parameter m is the number of training samples used. A perhaps unexpected phenomenon is observed here in that the curve has a maximum. This suggests that for a fixed number of training samples there is an optimal measurement complexity. Too many spectral bands or too many brightness levels per spectral band are undesirable from the standpoint of expected classification accuracy.

While at first surprising, the occurrence of this phenomenon is predictable as follows. One naturally would assume, as the upper left of Figure 3-22 indicates that as the dimensionality increases, the separability of classes would increase. On the other hand, as indicated by the upper right curves of Figure 3-22, one could have only a fixed finite number of samples by which to estimate the statistics that govern the classes. As one increases the measurement complexity by increasing the number of spectral bands, for example, a higher and higher dimensional set of statistics must be estimated with a fixed number of samples. Since one is endeavoring to derive a continually increasing amount of information detail from a fixed amount of data (the training samples) the accuracy of estimation must eventually begin to decrease. One could not expect good classifier performance if 10 dimensional statistics were to be estimated from 5 training samples, for example. Further, for a larger number of training samples, one would expect this performance decline to occur at a higher dimensionality. Taken together, these two effects would result in a performance curve as in the lower curve of Figure 3-22, a characteristic very similar to that Hughes was able to derive formally. The key observation to be made at this point is that there is an optimal value of measurement complexity in any given practical circumstance and more features do not necessarily lead to better results.

Though Hughes results were obtained on a theoretical basis, this phenomenon is commonly observable in practice.[16,17] For example, Figure 3-23 shows accuracy vs. number of features used for a 10-

134

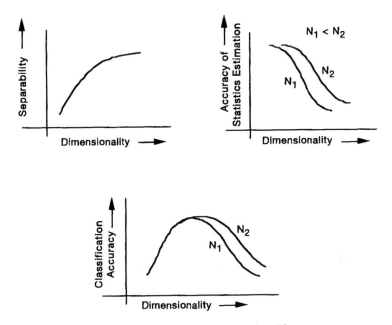

Figure 3-22. Effects thatresult in the Hughes Phenomenon.

class problem and 12-band data. The results shown are for a very large training set (11,000+ training samples), a moderate sized training set (1400+ samples), and a small set (228 samples) for the 10 classes. The training samples were drawn from a successively smaller portion of the same fields in each case, and the accuracy was determined from a much larger set of (70,635) test samples. It is seen that for the very large training set, the accuracy rises first

---

16 K. S. Fu, D. A. Landgrebe, and T. L. Phillips, Information processing of remotely sensed agricultural data, *Proceedings of the IEEE,* Vol. 57, No. 4, pp 639-653, April, 1969.

17 B. M. Shahshahani and D.A. Landgrebe, The effect of unlabeled samples in reducing the small sample size problem and mitigating the hughes phenomenon, *IEEE Transactions on Geoscience and Remote Sensing,* Vol. 32, No. 5, pp 1087-1095, September 1994.

The Hughes Effect

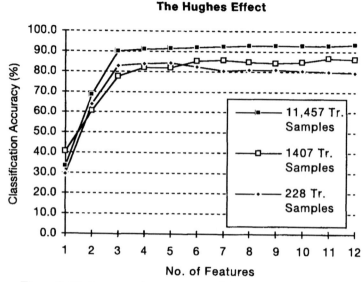

Figure 3-23. An example occurrence of the Hughes phenomenon.

very rapidly and then more slowly but steadily. The result for the moderate sized training set is similar but not quite so consistent. For the smaller training set, however, the accuracy reaches a peak of 84.6% for five features, then begins a slow decline to 79.5% at 12 features. Given that the number of training samples averaged only about 20 per class in this latter case, if even more features had been available, one might expect a more rapid decline in accuracy as the number of features began to equal, then exceed, the number of training samples.

It is further noted that there is a relationship between the number of training samples needed and the complexity of the classifier to be used. Fukunaga[18] proves under one example circumstance that the required number of training samples is linearly related to the dimensionality for a linear classifier and to the square of the dimensionality for a quadratic classifier. That fact is particularly relevant, since experiments have demonstrated that there are

136

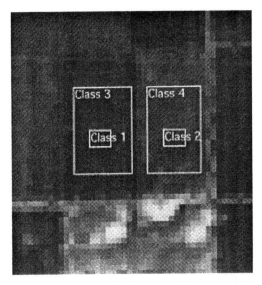

Figure 3-24. A small portion of a Landsat TM scene with test areas
for two agricultural fields. (In color on CD)

circumstances where second-order statistics are more appropriate
than first-order statistics in discriminating among classes in high
dimensional data.[19] In terms of nonparametric classifiers the
situation is even more critical. It has been estimated that as the
number of dimensions increases, the sample size needs to increase
exponentially in order to have an effective estimate of multivariate
densities.[20, 21]

[18] Fukunaga, K. *Introduction to Statistical Pattern Recognition.*
Academic Press, 1990.
[19] Chulhee Lee and D. A. Landgrebe, Analyzing high dimensional
multispectral data, *IEEE Transactions on Geoscience and Remote
Sensing,* Vol. 31, No. 4, pp. 792-800, July, 1993.
[20] Scott, D. W. *Multivariate Density Estimation.* Wiley, 1992.
[21] Hwang, J., Lay, S., Lippman, A., Nonparametric multivariate density
estimation: A comparative study, *IEEE Transactions on Signal
Processing,* Vol. 42, No. 10, pp. 2795-2810,1994.

A simple example may help to make the situation clearer. Figure 3-24 shows a small portion of a Landsat TM frame over an agricultural area with test areas for two agricultural fields marked. The image has been magnified so that the individual pixels are apparent. If the areas marked Class 1 and Class 2, each containing 12 pixels, were to be used as training areas for those classes, and it was proposed to use Thematic Mapper bands 1 and 3 for the classification, the estimated mean values and covariance matrices would be

$$\mu_1 = \begin{bmatrix} 83.4 \\ 25.7 \end{bmatrix} \qquad\qquad \mu_2 = \begin{bmatrix} 85.2 \\ 29.3 \end{bmatrix}$$

$$\Sigma_1 = \begin{bmatrix} 1.17 & \\ 0.06 & 0.24 \end{bmatrix} \qquad \Sigma_2 = \begin{bmatrix} 1.66 & \\ 0.73 & 2.97 \end{bmatrix}$$

$$\rho_1 = 0.11 \qquad\qquad \rho_2 = 0.33$$

Figure 3-25 shows a plot of the 12 data points, showing band 1 (abscissa) vs. band 3 (ordinate). In addition the area of concentration is shown for each class for a Gaussian density with the same mean vector and covariance matrix as the training data.

A normalized form of the covariance element may be computed as,

$$\rho_{13} = \frac{\sigma_{13}}{\sqrt{\sigma_1^2 \sigma_3^2}}$$

This quantity is the correlation coefficient of the data between the two bands. A correlation coefficient can vary over a range $-1 \leq \rho_{jk} \leq +1$, indicating the degree to which data in the two bands tend to vary in the same direction (positive correlation) or in opposite

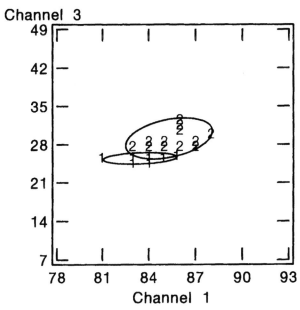

Figure 3-25. Scatter plot of the training samples for Classes 1 and 2. The ovals show the area of concentration for a Gaussian density with the same mean vector and covariance matrix.

directions (negative correlation). This turns out to be very useful information for a classifier, as seen in the example at the close of Chapter 1. Two bands that have positive correlation tend to be distributed along a line slanted at 45 degrees upward to the right. The higher the correlation, the more closely the data approach the line. The distribution is similar for negative correlation, but along a line upward to the left. Correlation values near zero imply distributions that tend to be circularly distributed, not having any favored direction.

Correlation between bands may in some instances seem undesirable. For example, it can suggest redundancy. However, and as previously noted, in the case of classification, another more positive

139

interpretation is appropriate. The mean value of a class defines where the class distribution is located in the feature space. The covariance matrix provides information about the shape of the distribution. Here it is seen that the higher the correlation between features, the more concentrated the distribution is about a 45 degree line. Zero correlation means it is circularly distributed, thus occupying a greater area (volume) in the feature space. Then there may be a greater likelihood that the distribution will overlap with a neighboring one. As an example of this, note the area of the circles in Figure 3-16 as compared to the area of the two class distributions occupied by the classes with two highly correlated features depicted there.

With this in mind, let us return to the consideration of the effect of the size of the training set. Consider defining the same user classes, but with a larger number of training samples. If instead of the areas marked Class 1 and Class 2 in Figure 3-24, the areas marked Class 3 and Class 4, each containing 200 points, are used as training samples, the corresponding results would be (Figure 3-26)

$$\mu_3 = \begin{bmatrix} 83.5 \\ 26.2 \end{bmatrix} \qquad\qquad \mu_4 = \begin{bmatrix} 86.9 \\ 31.2 \end{bmatrix}$$

$$\Sigma_3 = \begin{bmatrix} 1.86 & \\ 0.13 & 1.00 \end{bmatrix} \qquad \Sigma_4 = \begin{bmatrix} 3.31 & \\ 2.42 & 4.43 \end{bmatrix}$$

$$\rho_3 = 0.09 \qquad\qquad \rho_4 = 0.63$$

A comparison of μ_1 with μ_3 and μ_2 with μ_4, shows that there is relatively little change, indicating that the locations of the two distributions were reasonably well determined by the smaller training sets. However, the change in the corresponding covariance

140

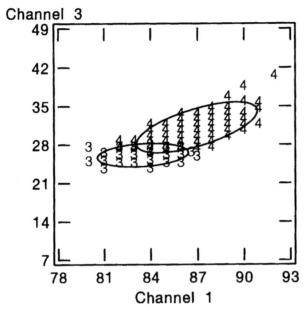

Figure 3-26. A scatter plot of the training samples for Classes 3
and 4 of the areas in Figure 3-25. The ovals again show the
area of concentration for a Gaussian density with the same
mean vector and covariance matrix as the data.

matrices is greater. The implication of this change is that the shape
of the distributions was not as well determined by the smaller
training sets.

This is a well-known result. The mean vector is known as a first-
order statistic, because it involves only one variable. The covariance
matrix is called a second-order statistic, because it involves the
relationship between two variables; the correlation shows how two
variables relate to one another. Higher-order statistics involve the
relationships between more variables. To perfectly describe an
arbitrary class density function would require knowing the value
of statistics of all orders. However, this would require an infinite
number of samples by which to estimate the statistics of all orders.

141

It is also the case that as the order of the statistic grows, the estimation process using a finite number of samples becomes more problematic. This is why one would in general expect that the mean vector would be reasonably well estimated with a smaller number of samples than would the covariance matrix, the circumstance we observed in the above example.

3.12 Summary to this Point

We have seen an approach to the analysis of multispectral data involving the use of pattern recognition technology:

- We began with a sensor system which serves as the Receptor by measuring the spectral response of each pixel, one at a time, in each of n spectral bands. The n-dimensional vector resulting is seen in an n-dimensional space, with the pixels of each class forming a distribution with a specific location and shape in that space.

- We saw how to design the Classifier portion of the pattern recognition system based on finding optimal discriminant functions.

- We examined three strategies for use in defining the discriminant functions, each requiring successively more information. The Maximum Likelihood scheme required only the class conditional density functions for the desired classes. The second, the Maximum A posteriori probability of correct scheme added to that the need for the prior probabilities of each of the classes. The third, the Minimum Risk scheme additionally required specification of the risk or loss associated with each possible decision.

142

- We next began to take notice of a circumstance that is nearly always present in the practical use of this technology, that of the usual limitation on the number of training samples available to determine the discriminant functions. We began by looking at the Gaussian density as a convenient density form by which to approximate the true class density functions, pointing out that a linear mixture of such density functions could be used to model complex shapes. A key advantage is that only parameters, and not the entire density function, must be estimated in this case. We briefly outlined other types of classifiers in order to provide some perspective at this point. Finally, we looked at an example one-dimensional estimation problem involving the histograms of grades on a test to begin to see on a conceptual but qualitative basis how the limited number of samples affects the conclusions one can draw about a density function. We concluded with the Hughes phenomenon, which shows more quantitatively the relationship between the number training samples available and the complexity of the classifier we choose to use.

Having now seen the relationship between finite training sets and the number of spectral features used, this raises the need to have a means for finding optimal, smaller subsets of features to use in classification. Not only will the use of less than all of the available features result in the need for less intense computation, but it could well lead to higher accuracy. Thus we next examine ways to predict accuracy and, from it, means for finding optimal subsets of spectral features to use.

3.13 Evaluating the Classifier: Probability of Error
Rarely in pattern recognition (or elsewhere!) is zero error possible, so classification error is an important system evaluation parameter.

In evaluating the three classifier strategies above, we saw briefly one of its uses and how it is done. We wish now to take a more in-depth look at evaluating and indeed projecting classification error.

An important and necessary element of the technology of pattern recognition is the ability to estimate the expected error rate after training but both before and after classification. In multispectral remote sensing, the uses of such an ability are many. It is of course needed to be able to provide a quantitative measure of how well a classifier will perform. Other uses include as an indicator of the adequacy of the training samples, the appropriateness of the class/subclass structure, and what features to select.

The means useful in evaluating the probability of error vary widely with the situation. In the case of discrete measurements, what is needed is called the "total probability" of error: pr(error) = Σpr(error, ω_i) summed over all classes, that is, the sum of the probabilities of the ways an error can occur. Thus one must list all the outcomes that can be in error,and then form the sum above. However,

$$pr(error, \omega_i) = pr(error \mid \omega_i)\, pr(\omega_i)$$

Therefore it is often convenient to use the RHS terms of this equation instead of those on the LHS, because these data are more readily perceivable and available. For example, in the maximum likelihood funny dice case, this is

$$pr(error) = pr(error|std)\, pr(std) + pr(error|aug)\, pr(aug)$$

$$= [\, pr(10|std) + pr(11|std) + pr(12|std)]\ pr(std) +$$

$$[\, pr(6|aug) + pr(7|aug) + pr(8|aug) + pr(9|aug)]\, pr(aug)$$

144

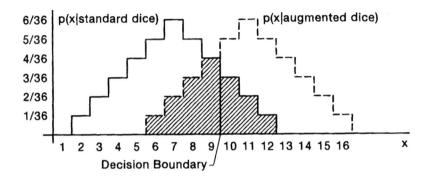

Figure 3-27. Probability of error determination for the funny dice problem.

Graphically, this is just the area under the discriminant function curves that is on the "wrong" side of the decision boundaries. See Figure 3-27.

In the case of a continuous outcome space, it is the corresponding area, but integration instead of summation is required to obtain it. For example, for a two class case,

$$Pr(error) = \int_{E_1} p(x \mid \omega_1)Pr(\omega_1)dx + \int_{E_2} p(x \mid \omega_2)Pr(\omega_2)dx$$

where E_1 and E_2 are the error regions of class 1 and class 2, respectively. In higher dimensional cases, the area becomes a volume. For example, Figure 3-28 shows the case conceptually for two dimensions. As can be seen, the determination of the limits on the needed integrals becomes complicated. When there are more than two classes or more than two dimensions, this evaluation becomes even more complicated.

145

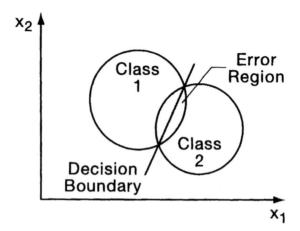

Figure 3-28. Probability of error evaluation in two dimensions.

Thus in practice error rates can be difficult to determine by this means. A direct calculation is usually not possible because, even in the parametric case, the integrals above can seldom be evaluated analytically. An indirect means is usually necessary.

There are a number of tools that have been developed to estimate classification error in practical circumstances.

1. Estimation from training samples. This method is sometimes referred to as the "resubstitution method." It is a matter of simply determining the error rate among the training samples and is usually an optimistic estimator. Also this approach is not particularly helpful when the purpose is to tell if the training of the classifier has been adequate.

2. Estimation from "test data." One can divide the pre-labeled data into two parts: use one part for training, and the other for test. Both parts must be representative of the class over the entire area.

146

3. The "leave one out" method. To use this method, one training sample is removed from the training set, the class statistics are estimated without it, the left-out sample is classified and its accuracy noted. The first sample is returned, a different one is withdrawn, and the process is repeated. The process is repeated until all samples have been used, and then the average classification accuracy is determined. This method is known to be an unbiased estimator of accuracy, but only to the extent that the training samples truly represent the class of interest.

4. Indirect estimators of classification error. This method is quite commonly used and is more fully described in the next section.

Statistical Separability

What is needed is a way to measure the degree of overlap, or alternately, the degree of separation of two probability

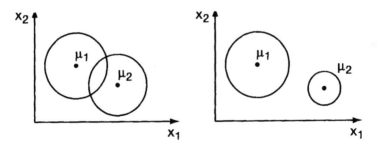

Figure 3-29. Separability in feature space showing dependancy on means and variances.

distributions.[22, 23] Initial perceptions about this might suggest metrics for this as follows:

1. The degree of separation is directly related to the distance between means, and inversely to the size of the variances. See Figure 3-29.

2. Thus one might consider using,

$$d_{norm} = \frac{|\mu_1 - \mu_2|}{\sigma_1 + \sigma_2}$$

However, this measure does not have quite the correct properties. For example, it is zero for equal means even when the variances are not equal. Just looking at mean differences, whether or not normalized by the variance does not provide what is needed.

3. A bit more sophisticated look at the problem is needed. See Figure 3-30. Clearly, the larger a, the better. The ratio of likelihood values,

$$L_{ij} = \frac{p(\mathbf{X} \mid \omega_i)}{p(\mathbf{X} \mid \omega_j)}$$

is directly related to a. The log of the likelihood ratio,

---

[22] P. H. Swain, and S.M. Davis, editors, *Remote Sensing: The Quantitative Approach*, McGraw-Hill, 1968 p164 ff

[23] J. A. Richards, and Xiuping Jia, *Remote Sensing Digital Image Analysis*, 3rd edition, Springer-Verlag, 1999, Chatper 10.

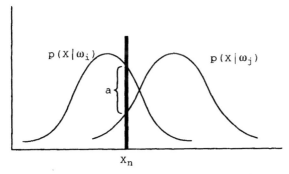

Figure 3-30. Two class density functions with a means for measuring how they differ.

$$L'_{ij}(X) = \ln L_{ij}(X) = \ln p(X \mid \omega_i) - \ln p(X \mid \omega_j)$$

is directly related also and is mathematically more convenient. The quantity Divergence, D_{ij}, of classes i and j is defined based on this, as,

$$D_{ij} = E[L'_{ij}(X) \mid \omega_i] + E[L'_{ji}(X) \mid \omega_j]$$

Where E[.] is the mathematical expectation of the random quantity contained in the brackets. Properties that make it useful may be summarized as follows:

- It is always positive,

- D_{ij} between a function and itself is zero,

- It is symmetrical ($D_{ij} = D_{ji}$),

- If the vector components are independent, the divergence in a multivariate case is the sum of the univariate components,

149

- The Divergence of classes in a subspace is always less than that in the full space,

- It is relatively easy to compute in the Gaussian case. The expression for Divergence in this case is,

$$D = \frac{1}{2} tr\left[\Sigma_i - \Sigma_j \right]\left[\Sigma_i^{-1} - \Sigma_j^{-1} \right] + \frac{1}{2} tr\left[\Sigma_i^{-1} + \Sigma_j^{-1} \right]\left[\mu_i - \mu_j \right]\left[\mu_i - \mu_j \right]^T$$

An Example Use of Statistical Separability

To illustrate how a statistical separability measure such as Divergence might be used, consider the problem of feature selection. Feature selection is the problem of selecting the best subset of features from a larger set, to use in a specific classification. As seen from the Hughes phenomenon, depending on the size of the training set, higher classification accuracy may be obtained by using a smaller set of features. Suppose that it is desired to find the best four features to use for the classification of 12 band data of an

Classes and Symbol		Spectral Bands
Corn	C	1: 0.40-0.44 μm
Soybeans	S	2: 0.44-0.46 μm
Wheat	W	3: 0.46-0.48 μm
Alfalfa	A	4: 0.48-0.50 μm
Red Clover	R	5: 0.50-0.52 μm
Rye	Y	6: 0.52-0.55 μm
Oats	O	7: 0.55-0.58 μm
Bare Soil	X	8: 0.58-0.62 μm
		9: 0.62-0.66 μm
		10:0.66-0.72 μm
		11: 0.72-0.80 μm
		12: 0.80-1.00 μm

Table 3-3. Classes, symbols, and spectral bands for an example agricultural crops classification.

agricultural crops area. Table 3-3 shows the classes, symbols, and spectral bands that apply.

Table 3-4 shows a portion of the output from a typical feature selection calculation. The calculation proceeds by first developing a list of all of possible 4-tuples of the 12 features. Five of these are listed in the table under Spectral Bands. Next, the statistical separability is calculated between each possible pair of classes for each 4-tuple, listed under Individual Class Separability, using the symbols given in Table 3-3. Following this, the average interclass distance of each class pair of each 4-tuple is determined and the data for each such 4-tuple, listed in descending order of average interclass distance.

The 4-tuple 1,9,11,12 is seen to have the highest average separability, based on the particular distance measure used. Thus this is a candidate to be considered for a four band classification.

Class Separability Measure
(No Maximum)

Rank	Spectral Bands	Average Separability	Individual Class Separability				
			SC	SW	SA	WR	WY
		• • •					
1	1, 9, 11, 12	444	25	190	188	620	58
2	6, 9, 11, 12	428	26	177	229	630	208
3	2, 9, 11, 12	423	24	151	182	619	58
4	5, 9, 11, 12	420			•		
5	8, 9, 11, 12	•			•		
	•	•			•		

Table 3-4. The results of a feature selection calculation for determining the best 4 bands out of 12 to use in classification.

151

However, note that the class pair WR (Wheat and Red Clover) has a much larger separability measure than several of the others, indicating that these two classes are very separable and the large statistical separability resulting may be excessively influencing the determination of the average. In Table 3-4 it is seen that discriminnating between Soybeans and Corn (SC) may be a much more difficult task. Thus, as a practical matter, one may wish to choose a 4-tuple indicating a higher separability for this class pair even though it might have a somewhat lower overall average. For this reason, one may want to scan the entire table of Individual Class Separability values before making a final choice of which 4-tuple of bands to use. This illustrates well how a statistical separability measure can be very useful in making the final decision on the setting of parameters for a classification in a multiclass case.

In a high dimensional case, another shortcoming of feature selection methods can be the computational load. To select an m feature subset out of a total of n features, the number N of combinations that must be calculated is given by the binomial coefficient,

$$ N = \binom{n}{m} = \frac{n!}{(n-m)!\,m!} $$

This number can become quite large. For example, to find the best 10 features out of 20, there are 184,756 combinations that must be evaluated. For the best 10 out of 100, the number of combinations exceeds 1.7×10^{13}. There is really no simple direct solution to this dilemma. Several suboptimal methods are sometimes used, such as a sequential forward or sequential backward search.

In such higher dimensional cases, another possibility is a feature space transformation–referred to as Feature Extraction–such as the

Discriminant Analysis Feature Extraction or Decision Boundary Feature Extraction. These will be discussed in Chapter 6.

Characteristics of Statistical Separability Measures

As indicated above, statistical separability measures have a number of uses in practical pattern recognition problems. How one chooses to use a particular statistical separability measure in a given situation depends on the specific circumstances. For this reason, it is useful to know the general characteristics of several distance measures. For example, Divergence has an advantage in that it is a shorter calculation than some. However, it has a significantly nonunique, nonlinear mapping from D_{ij} to pr(error). Figure 3-31 shows the relationship between Divergence and classification accuracy. Note that a divergence value of D_1 could imply any value of accuracy between p_1 and p_2. Note also that a significantly different divergence value, either higher or lower could imply accuracies included in that same range. Further, there appears to be a significant saturation

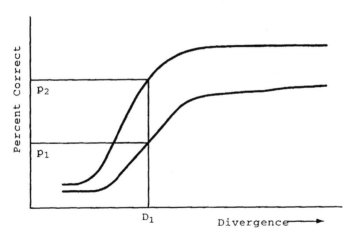

Figure 3-31. Graph showing the nonunique, nonlinear relationship between Divergence and the error rate or percentage correct that a specific value of Divergence implies.

effect, whereby little difference in percent correct is indicated in the high accuracy areas for a large difference in Divergence values.

Divergence was one of the earliest statistical separability measures used by the remote sensing community, but, as just indicated, it does have significant shortcomings. There were several attempts to minimize these shortcomings. For example, to reduce the nonlinear nature of the relationship between Divergence and classification accuracy, a transformation, $D^T_{ij} = 2[1 - \exp(- D_{ij}/8)]$ was proposed and used for a period of time. This did seem to help. Figure 3-32 shows the result of a Monte Carlo test of this relationship by randomly generating two-class, two-dimensional sets of data with different means and covariance matrices, calculating the Divergence between the two classes, then classifying the data and noting the accuracy. It is seen that the relationship is indeed somewhat more linear, but not one-to-one, especially in the higher accuracy range.

Table 3-5 provides a list of a number of the statistical separability measures that have appeared in the literature. That there are so many entries suggests the degree to which this problem has been studied, and at the same time suggests the importance of studying the characteristics of a distance measure for any particular use. In the case of Gaussian densities, several of the above may be computed as shown in Table 3-6.

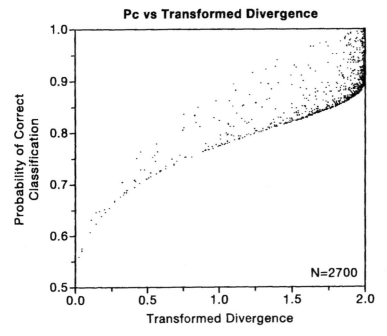

Figure 3-32. The results of a Monte Carlo test of the relationship between probability of correct classification of two classes in two-dimensional space vs. Transformed Divergence.

Name	Mathematical Form
Divergence	$D = \int_{-\infty}^{\infty} \ln \dfrac{f(x)}{g(x)} [f(x) - g(x)] dx$
Bhattacharyya	$B = -\ln \int_{-\infty}^{\infty} \sqrt{f(x)g(x)}\, dx$
Jeffreys-Matusita	$J = \sqrt{\int_{-\infty}^{\infty} \left\{ \sqrt{g(x)} - \sqrt{f(x)} \right\}^2 dx}$
Cramer-Van Mises	$W = \sqrt{\int_{-\infty}^{\infty} [G(x) - F(x)]^2 dx}$
Kiefer-Wolfowitz	$V = \int_{-\infty}^{\infty} |F(x) - G(x)| \varepsilon^{-|x|} dx$
Kolmogorov-Smirnov	$K = \sup_x |G(x) - F(x)|$
Kolmogorov Variational	$K(p) = \int_{-\infty}^{\infty} |p_g g(x) - p_f (f(x)| dx$
Kullback-Liebler Numbers	$L_{fg} = \int_{-\infty}^{\infty} \ln\!\left(\dfrac{f(x)}{g(x)} \right) f(x) dx$
Mahalanobis	$\Delta = \sqrt{(\mu_f - \mu_g)^T \Sigma^{-1} (\mu_f - \mu_g)}$
Samuels-Bachi	$U = \sqrt{\int_0^1 \{F^{-1}(\alpha) - G^{-1}(\alpha)\} d\alpha}$

where $F^{-1}(\alpha) = \operatorname{Inf}\{c \mid Q_c \ Q_\alpha \neq 0\}$

and $Q_c = \left\{ x \mid \sum_{i=1}^{q} x_i \leq C \right\}$, $\quad Q_\alpha = \{x \mid F(x) \geq \alpha\}$

Table 3-5. Some statistical separability measures that have appeared in the literature.

Name   *Mathematical expression*

Divergence

$$D = \frac{1}{2} \text{tr}\left[(\Sigma_f - \Sigma_g)(\Sigma_f^{-1} - \Sigma_g^{-1})\right] + \frac{1}{2} \text{tr}\left[(\Sigma_f^{-1} - \Sigma_g^{-1})(\mu_i - \mu_j)(\mu_i - \mu_j)^T\right]$$

Bhattacharyya

$$B = \frac{1}{8}[\mu_f - \mu_g]^T \left[\frac{\Sigma_f + \Sigma_g}{2}\right]^{-1}[\mu_f - \mu_g] + \frac{1}{2}\ln\frac{\left|\frac{1}{2}[\Sigma_f + \Sigma_g]\right|}{\sqrt{|\Sigma_f||\Sigma_g|}}$$

Jeffreys-Matusita

$$J = \sqrt{2\left\{1 - \frac{\sqrt[4]{|\Sigma_f||\Sigma_g|}}{\sqrt{\frac{1}{2}|\Sigma_f + \Sigma_g|}}\exp\left\{-\frac{1}{8}[\mu_f - \mu_g]^T\left[\frac{\Sigma_f + \Sigma_g}{2}\right]^{-1}[\mu_f - \mu_g]\right\}\right\}}$$

Kullback-Liebler Numbers

$$L_{fg} = \frac{1}{2}\text{Ln}\frac{|\Sigma_f|}{|\Sigma_g|} + \frac{1}{2}\text{tr}\Sigma_f\left[\Sigma_g^{-1} - \Sigma_f^{-1}\right] + \frac{1}{2}\text{tr}\Sigma_g^{-1}[\mu_f - \mu_g][\mu_f - \mu_g]^T$$

Mahalanobis

$$\Delta = \sqrt{[\mu_f - \mu_g]^T \Sigma^{-1}[\mu_f - \mu_g]}, \quad (\Sigma = \Sigma_f = \Sigma_g)$$

Swain-Fu

$$T = \frac{|\mu_g - \mu_f|}{D_f + D_g} \quad \text{where} \quad D = \sqrt{\frac{|\mu_f - \mu_g|^2(q+2)}{\text{tr}\left\{\Sigma^{-1}(\mu_f - \mu_g)(\mu_f - \mu_g)^T\right\}}}$$

Table 3-6. The Gaussian form for several distance measures.

157

Unfortunately, the relationship between distance measures and classification accuracy is not precisely one-to-one for any of the measures above, meaning that a given value of a distance measure does not imply a specific value of probability of correct classification. Rather, the best that can be said is that (1) a given value of a distance measure implies a certain range of possible classification accuracies, and (2) usually (but not always), a larger value of a distance measure implies a larger value of classification accuracy. This is an important limitation, and has been studied in the literature extensively. Only in a few cases has it been possible to derive bounds on the probability of correct classification. One such is for the Jeffries-Matusita distance where the bounds are found to be

$$\frac{1}{16}\left(2 - J_{ij}\right)^2 \leq Pe \leq \frac{1}{4}\left(2 - J_{ij}\right)$$

where Pe is the probability for error. A graph of the bounds on Probability of Correct Classification (= 1 − Pe) for the Jeffries-Matusita distance is given in Figure 3-33.

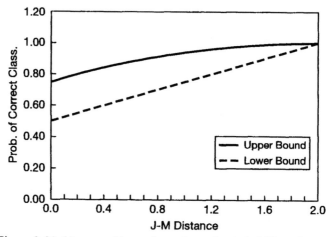

Figure 3-33. Upper and lower bounds on the probability of correct classification vs. JM distance.

It may be shown that in the Gaussian case, the relationship between the Jeffries-Matusita distance and Bhattacharyya distance is given by $J_{ij} = 2(1 - e^{-B_{ij}})$, where B_{ij} is the Bhattacharyya distance. Thus bounds for the Bhattacharyya distance case can also be obtained. In cases where it has not been possible to derive such error bounds analytically, it is useful to study the property of separability measures empirically. In the graphs of Figure 3-34 are shown graphically the results of Monte Carlo studies of the relationship between several distance measures and classification accuracy. The results that fall in a narrower range imply a greater degree of unique one-to-one mapping between the distance measure and classification accuracy. It is seen that some metrics imply a relationship that is much more nearly linear than others. Note also how the use of the Gaussian error function transformation,

$$\phi(x) = \frac{1}{\sqrt{2\pi}} \int_{-\infty}^{x} \varepsilon^{t^2/2} dt$$

indicated by ERF, tends to linearize the relationship.

From these results for two-dimensional data, the Error Function Bhattacharrya Distance appears to most nearly provide the kind of performance desired, with direct Bhattacharrya Distance not far behind. Given the larger amount of computation required for the error function transformation, the direct Bhattacharrya Distance measure may be a good choice for many circumstances.

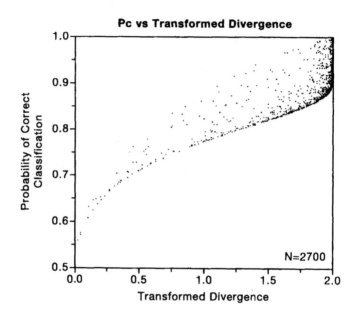

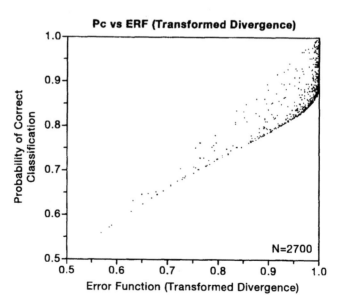

Pc vs Chernov Divergence

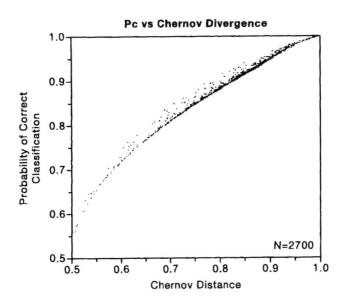

Pc vs ERF(Chernov Divergence)

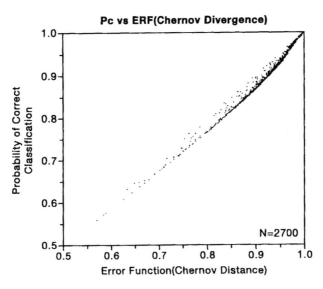

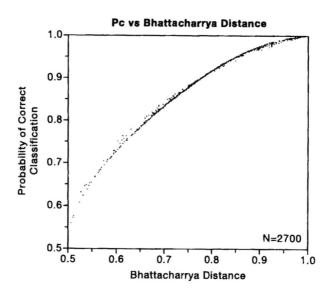

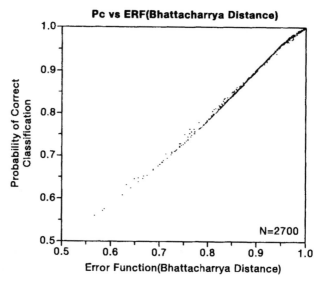

Figure 3-34. Results of a Monte Carlo study of the relationship between Probability of Correct Classification and several different statistical separability measures.

3.14 Clustering: Unsupervised Analysis

Another type of classification is called unsupervised classification, or simply clustering. It is referred to as unsupervised because it does not use training samples, only an arbitrarily number of initial "cluster centers" which may be user-specified or may be quite arbitrarily selected. During the processing, each pixel is associated with one of the cluster centers based upon a similarity criterion.

Clustering algorithms are most commonly used as an aid to selecting a class list and training samples for the classes in that list. Fundamentally, to be optimally useful, a classification must have classes that are (simultaneously)

- Of information value,
- Exhaustive, and
- Separable

The training samples generally are selected with emphasis on the former one. Clustering is a useful tool of the training process to achieve the latter two. Rarely in remote sensing are classes so separable that an unsupervised algorithm will lead to an acceptable final classification by itself. It can be a useful procedure, though, in defining spectral classes and training for them by breaking up the distribution of pixels in feature space into subunits so that one can observe what is likely to be separable from what. It allows one to locate the prevailing modes in the feature space, if any prevalence exists. These could become the subclasses in the case of a quadratic classifier, for example. The clustering concept will be illustrated in terms of the hypothetical data shown in Figure 3-35.

In Figure 3-35, one can see what appear to be two rather obvious clusters of data, which in this case are easily associated with two classes of informational value. To develop a clustering algorithm

163

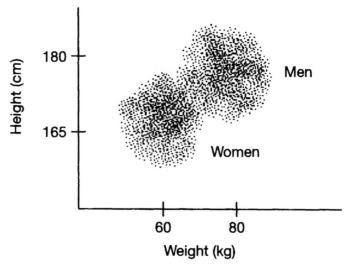

Figure 3-35. Hypothetical two-dimensional data suggesting clusters.

that might divide the data into these two classes, three basic capabilities are needed.

1. A measure of distance between points,
2. A measure of distance between point sets, and
3. A clustering criterion.

One uses the points distance measure to decide which points are close to one another to form point sets (i.e. clusters) and the point sets distance measure to determine that clusters are sufficiently distinct from neighboring point sets. The clustering criterion is used to determine if each of the clusters is sufficiently compact and also adequately separable from the other clusters.

1. Example points distance measures:

Euclidean Distance:

$$d(\mathbf{x}_1, \mathbf{x}_2) = \|\mathbf{x}_1 - \mathbf{x}_2\|$$

$$= \{(\mathbf{x}_1 - \mathbf{x}_2)^T (\mathbf{x}_1 - \mathbf{x}_2)\}^{1/2}$$

$$= \{\sum_{i=1}^{N} (x_{1i}^2 - x_{2i}^2)\}^{1/2}$$

or, L_1 distance:

$$d(\mathbf{x}_1, \mathbf{x}_2) = \sum_{i=1}^{N} |x_{1i} - x_{2i}|$$

2. A measure between point sets, whereby any of the previously described statistical separability measures might be used, but more frequently simpler measures are used.

3. A common clustering criterion or quality indicator is the sum of squared error (SSE),

$$SSE = \sum^{\omega_i} \sum^{x \in \omega_i} (\mathbf{x} - \mu_i)^T (\mathbf{x} - \mu_i)$$

where μ_i is the mean of the ith cluster and $\mathbf{x} \in \omega_i$ is a pattern assigned to that cluster. The outer sum is over all the clusters. This measure computes the cumulative distance of each sample from its cluster center for each cluster individually and then sums these measures over all the clusters. If it is small, the distances from patterns to cluster means are all small and the clustering would be regarded as satisfactory.

Typical steps for a clustering algorithm might be as follows.
Step 1. Initialize: Estimate the number of clusters needed, and select an initial cluster center for each. The user may select

165

the initial points, a measure of supervision, or the points be arbitrarily selected.

Step 2. Assign each vector to the nearest cluster center, using criterion 1 above.

Step 3. Compute the new center of each cluster, and compare with the previous location. If any have moved, return to step 2.

Step 4. Determine if the clusters have the characteristics required.

• Are they sufficiently compact, using the inner sum of criterion 3 above? If not, subdivide any that are too distributed, assume cluster centers for the new ones, and go back to step 2.

• Are they sufficiently separated from other clusters, using criterion 2 above? If not, combine those that are too close together, assume a new cluster center and go back to step 2. If so, clustering is complete.

It is seen that the process is an iterative one, continuing until the criteria in step 4 are all met. Figure 3-36 illustrates in concept the migration of the clusters and their centers during the iteration. Even though the initial cluster centers were not well located in the first place, the iterative process eventually leads to a reasonable location for them, although it would, no doubt, need to pass through many more iterations than might have been necessary had the initial centers been better located. Notice that such a procedure has the ability to end up with either more or less cluster centers than it started with.

An algorithm that follows these steps is referred to as an isodata method (from the name given to this algorithm by its originators) or by the term *migrating means* algorithm. In general, the results produced by clustering algorithms are dependent, often critically

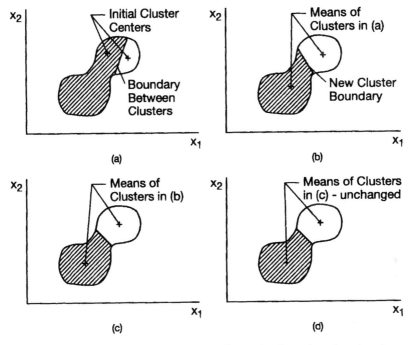

Figure 3-36. A hypothetical sequence of steps in clustering showing the migrating cluster centers.

so, on the choice of the number and location of initial cluster centers. With actual multispectral data, choosing different initial cluster centers usually results in quite different final clusters.

Initial clustering centers are commonly chosen by one of the following:

- Equally spaced along the major diagonal of the data set in feature space,

- By means of a simpler, noniterative clustering algorithm, or,

167

- Based on the mean value of user specified (training) samples.

Athough it is common to introduce the topic of clustering or unsupervised classification by means of sketches as above, where the data have rather obvious clusters, such sketches are misleading. Actual multispectral remote sensing data are typically distributed in a continuum over some portion of the feature space. Thus, it is unlikely that a clustering algorithm by itself can locate clusters that are specifically related to classes of interest. Clustering does not lead to a unique result; any change in clustering algorithm parameters or the algorithm itself is likely to lead to a different set of clusters. Thus, while clustering can be a very useful tool in defining classes and class statistics, it generally must be used in conjunction with additional information in some form as to what final classes are desired.

In the following is shown an example clustering similar to the conceptual one above only using real data. The data are from a 12-band data set gathered over an agricultural area. Figure 3-37 shows a simulated color IR presentation of the data in image space using three of the 12 bands. A sub-area marked by the rectangle is to be clustered using only two of the 12 bands. Figure 3-38 shows the data from the rectangular area in feature space. Though the rectangular area includes data from two different crop species, it is less clear in the two-band feature space which pixels belong to which species.

Figure 3-39 shows the location of the arbitrarily selected initial cluster centers as round dots. It also shows the final location of the cluster centers after nine iterations. Figure 3-40 shows in feature space how the pixels were assigned to the two clusters.

168

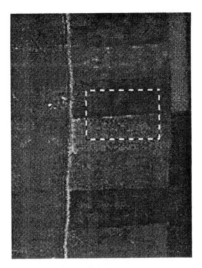

Figure 3-37. Simulated color IR
image of agricultural area with an
area marked for clustering. (In color
on CD)

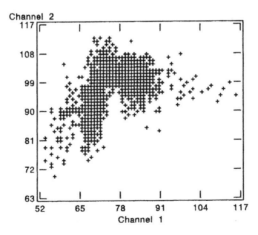

Fiure 3-38. Scatter plot of the data in two bands from the region marked in
Figure 3-37.

169

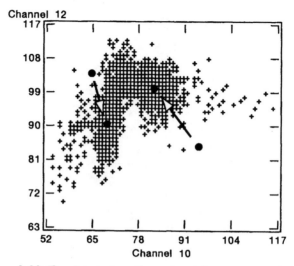

Figure 3-39. The data in two-dimensional feature space showing the initial, arbitrarily selected cluster centers (round dots), and their final location (diamond dots).

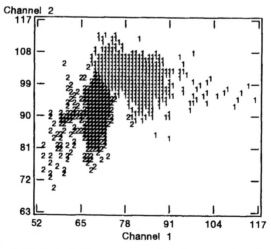

Figure 3-40. Data in feature space showing to which cluster each was associated.

170

Classes
■ Cluster 1
■ Cluster 2

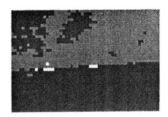

Figure 3-41. Thematic presentation in image space showing how
the pixels of Figures 3-37 and 3-38 were assigned to the two
classes as a result of the clustering. (In color on CD)

Figure 3-41 shows the assignment of pixels to clusters in image space. It is seen here that the assignment of pixels to clusters was only approximately what would be desired in order to have the clusters match the species classes. This is partly due to the fact that for this illustrative example, only two bands were used to make the feature space presentation a little easier to understand. However, as one moves to higher dimensional spaces, the data become increasingly sparse. This tends to provide more potentia for more supervision by the analyst to be supplied in order to obtain the specific desired result in any given case.

3.15 The Nature of Multispectral Data in Feature Space
Before looking specifically at the steps needed to analyze multispectral data, it may be useful to think for a bit about what the nature of data really is like in feature space and why. As has just been seen in the clustering example, it is usually an oversimplification to think that different classes of data just naturally occur in clusters that are fairly distinct in feature space, at least for classes of a typical degree of detail. The field in the upper portion of the area designated in Figure 3-37 is soybeans, while that in the lower portion is red clover. These are two carefully cultivated but quite different species of vegetation. Each might be expected to be much more uniform than, for example, natural vegetation, and yet,

171

even in two-dimensional space, they did not display as two distinct points. Rather, as Figure 3-38 showed, the data for each class occurred as a distribution of points. The distribution even for these relatively disparate species spread widely enough about their average or "typical" value that they overlapped to some degree, such that any nonsupervised clustering algorithm could not be expected to successfully divide them.

Why does this distribution occur? Why isn't the response from such a uniform crop area more uniform? Each pixel in that data covers an area several meters across. In an area of this size there are many different reflectance surfaces, each with its own spectral response, areas such as the cell structure of the leaf, leaf veins and stems, and perhaps even soil. In addition, depending on the sun angle and view angle for a given pixel, various of those reflectance surfaces will have different degrees of illumination from deep shadow to fully illuminated, even to the point that some surfaces will be reflecting in a specular (mirrorlike) fashion. The latter case occurs on small spots on leaf surfaces where the angles between the solar illumination, the leaf surface, and the view angle are just right. This angular relationship would occur only on small spots, but the reflected light would be very bright and would have a spectral distribution more due to the illumination source than the volumetric reflectance properties of the plant. Just how much of the total area of a pixel reflects in this specular fashion, a property of the physical structure of the plant in this case, and depending on the particular mixture of all of these reflectance surfaces and conditions that happens to occur in a given pixel, the net response measured from different pixels differs. Further, wind across the canopy would vary the geometry of the plants in different parts of the field in different ways, so that response would vary considerably.

The same sort of complexity is true of any Earth surface area, natural or human-made. For example, even for a one-centimeter pixel over

172

a desert with only a surface of sand, the sand granules could be expected to have a varying makeup of molecules, and the granular surface would have varying shadowed vs. illuminated surface effects. The slope and aspect of the soil area would also be a significant additional variable for the spectral response from pixel to pixel. This is the reason why the analysis of multispectral data is fundamentally a process of discriminating between distributions of spectral responses rather than individual spectral responses.

Another issue that need be kept in mind as one prepares to analyze data is the role of dimensionality in feature space. In the case of Figure 3-38, it is clear that two dimensions are not enough to successfully discriminate between these two plant species. What would happen if the very same pixels were now distributed over three-dimensional space, i.e., the two shown plus a third extending upward out of the paper? It is not difficult to visualize that the pixels would now be at least somewhat more separable. If that were not enough to accomplish what is desired, a fourth or more dimensions could be added, causing the same pixels to be distributed over the larger and large volume of the higher-dimensional space.

Yet another means for increasing the spread of pixels in feature space so they might be more separable might be to increase the bit precision of the data based on an improvement of the S/N. This would enlarge the volume of the space by providing the two axes with more discrete locations. This could also increase the possibility of separation between groups of pixels defining a class of interest.

But either of these ways of increasing the volume of the discrete-valued feature space, while potentially beneficial, would require greater precision in locating the decision boundaries in feature space to realize that potential. This is the message that the Hughes phenomenon results were delivering.

173

Summarizing then, one can expect multispectral data for classes of interest to occur in distributions, usually those that overlap to some degree. The distributions are not likely to be the same from time to time or place to place, because they are the result of the combination of many factors that will vary in their effect. One can expect to improve the separation between classes by increasing the number of bands and/or by increasing the S/N. However, doing so increases the needed precision with which the classes are modeled. These factors are important to keep in mind as one approaches the analysis of a data set.

3.16 Analyzing Data: Putting the Pieces Together

So far the discussion has focused on data characteristics and the various tools that might be used in the process of analyzing a given multispectral data set. Next is discussed how these tools might be put together to carry out an actual classification. Typical steps in analyzing a low or moderate dimensional multispectral data set[24] might be as follows:

1. **Preprocessing**
   - Refomat the data as needed
   - Examine the data for artifacts or other such problems.
   - Carry out any needed geometric & radiometric adjustment of the data.

2. **Class definition and training**
   - Apply clustering and separability analysis to discover modes in the data.
   - Select classes and subclasses, and training samples for each.

---

[24] For high-dimensional data, some of the processes listed would not be helpful, and others must be added. Analysis of high dimensional data will be more specifically treated in later chapters.

• Compute class statistics.

3. **Feature determination**
   • Use feature selection calculation to determine the best feature set.
   • Make preliminary determination of performance, iterate previous steps as necessary.

4. **Classification**
   • Select the type of algorithm to be used.
   • Classification
   • Make a further preliminary evaluation, re-iterate previous steps as necessary.

5. **Post classification processing**
   • Make final evaluation.
   • Extract information (summation).
   • Present information in map or tabular form.

There are many approaches to analyzing multispectral image data, almost as many as there are analysts. The steps used in any given analysis must necessarily be based upon the scene, the information desired, and the data analyst's initial assumptions about the data characteristics.

The basic requirement of an analysis process is

   • To divide the entire multidimensional feature space defined by the data set into an exhaustive set of nonoverlapping regions in such a way that the regions delineated, or an appropriate subset of them, correspond as precisely as possible to data points belonging to the particular classes desired.

175

Experience has shown that, if properly used, the assumption that each of these data subsets may be modeled in terms of one or a combination of Gaussian distributions is a quite practical and powerful way to proceed. Among the advantages of using the Gaussian model is the mitigation of need for large training sets to properly define the desired classes, especially where the spectral dimensionality is large. However, use of this assumption does impose on the analysis process, some means for identifying the various modes of each desired information class, and the fitting of densities to each of these modes. It would be naive to assume that a single Gaussian density would fit every class of interest. *The analysis process then consists of finding the parameters for a collection of Gaussian distributions that fit the entire data set, and for which some subset of the distributions correspond to the classes of interest.*

With regard to dealing with confounding scene observation variables such as the atmosphere, illumination and view angles, and terrain relief, one approach is to attempt to pre-adjust the data before analysis to account for the variation introduced by each of these confounding variables into spectral responses. However, as a practical matter, the lack of adequately precise values for the parameters of such scene variables, the complexity of the calculations required even if the parameter values are known, and the noise and uncertainties introduced by such calculations (which themselves can tend to reduce the amount of information that could ultimately be obtained from the data) are disadvantages of this approach.

A second approach is to use analysis techniques that are relatively insensitive to such scene observation variables in the first place. The various conditions in which a desired information class exists in the data may be handled by defining additional subclasses for

176

the desired information classes. This is the approach in mind in the following, and a minimum amount of pre-analysis adjustment to the data is assumed to have been used.

The steps below, then, allow for accomplishing the analysis task according to the above approach.

1. Familiarization with the data set.

    •   Display the data in image space.

    •   Compare the displayed image with any ground reference information about the site that may be available.

    •   View histograms of all or an appropriate subset of bands to check the quality of the data.

    •   Compose a tentative list of classes that is adequately (but not excessively) exhaustive for this data set.

2. Preliminary selection of the classes and their training sets.

    •   Using clustering, cluster the area from which training fields are to be selected, saving the results to disk file.

    •   Display the resulting thematic map for use in marking training areas.

    •   Using either the display of the original data or that of the thematic cluster map, make a preliminary selection of training fields that adequately represent the selected classes.

3. Verification of the class selection and training.

- Use the feature selection process to determine the degree of separability between the various classes.

- Check the modality of the classes by examining the cluster map or by clustering the training areas. Where multi-mode classes are found, it may be appropriate to define two or more sub-classes to accurately represent the entire class. It may be desirable to iterate between steps 2 and 3.

- In low-dimensional cases, it may be appropriate to examine the histograms of each class in a representative set of spectral bands to determine the need for subclasses. This is another means of identifying the need for subclasses.

4. Selection of the spectral features to be used.

- In the case of a data set with moderate dimensionality, once a reasonably final training set is reached, use the feature selection process to choose the best subset of features for carrying out the classification for a given training set.

- In the case of a data set with high dimensionality, once a reasonably final training set has been reached, use feature extraction to construct a set of optimal features. Feature extractions are special algorithms suited to high dimensional data. They are discussed in detail in Chapter 6.

5. Preliminary classification of the data.

- Classify the training fields only, using the spectral bands you have selected to verify their purity and separability.

6. Final classification, evaluation of the classification, and extraction of the desired information.

- Classify the entire data set using the features selected.

- Mark as many fields as possible as test fields, and use them to determine measures of the accuracy obtained on the test fields and to determine how well the classifier generalizes beyond the training set.

- Make modifications to the training as required to obtain satisfactory results at this point.

- Depending upon the results of these evaluations, it may be necessary to repeat previous steps after modifying the class definitions and training. After becoming satisfied with the results, classify the entire data set, perhaps setting a modest threshold value.

- Generate thematic map versions of the results and an image space map showing the degree of membership of each pixel to its assigned class for subjective evaluation purposes. The classification results display is useful in determining that the classification results are appropriate and consistent from a spatial distribution standpoint. The portion of points thresholded in the results display, together with the degree of membership thematic map help to determine if any important modes in the data have been missed in the class definition process.

- Depending on the outcome, it may again be necessary to iterate using some of the above steps.

- Determine a final quantitative evaluation of the results based upon the accuracy figures of the training and test fields classification.

A common means for quantifying the accuracy of a classification is to use the so-called Accuracy Matrix, sometimes called the Confusion Matrix. An example is as follows:

Class Name	Accur.+ (%)	No. of Samples	Number of Samples in the Class						
			Roof	Street	Path	Grass	Trees	Water	Shadow
Roof	98.8	15364	15176	96	91	0	0	0	1
Street	93.4	1812	75	1693	0	0	2	2	40
Path	99.5	429	2	0	427	0	0	0	0
Grass	99.9	5904	0	0	0	5900	4	0	0
Trees	99.4	1748	0	0	0	11	1737	0	0
Water	80.3	4791	383	3	0	0	7	3847	551
Shadow	87.5	279	14	19	0	0	1	1	244
TOTAL		30327	15650	1811	518	5911	1751	3850	836
Reliability Accuracy (%)*			97.0	93.5	82.4	99.8	99.2	99.9	29.2

OVERALL CLASS PERFORMANCE (29024 / 30327) = 95.7%
Kappa Statistic (X100) = 93.6%. Kappa Variance = 0.000003.
+ (100 - percent omission error); also called producer's accuracy.
* (100 - percent commission error); also called user's accuracy.

The rows in the table list the classes and indicate how the labeled pixels of each class were assigned to classes by the classifier. One can thus quickly see what kinds of errors were made, provided that the labeled samples were indeed correctly labeled. An accuracy value is listed for each class. This accuracy value is sometimes called the "Producer's Accuracy" and as a percentage is equal to 100% minus the percent omission error. Just below the line labeled "TOTAL" is a line labeled "Reliability Accuracy." This one is sometimes called the "User's Accuracy" and as a percentage is equal to 100% minus the commission error.

The overall percent accuracy, which is the total number correctly classified divided by the total number classified (i.e. the sum of the observations on the diagonal of the accuracy matrix divided by the total number of observations) is perhaps the simplest means for assessing classification accuracy by a single number. However, it does not take into account the degree of balance in the errors. That is, even if all of the pixels were not assigned to the correct class, it would be desirable for the proper proportion of the pixels be assigned to each class.

Another popular means for accessing accuracy is the so-called Kappa statistic.[25] A means for estimating this statistic, as given by Congalton[26] is as follows:

$$\hat{K} = \frac{N\sum_{i=1}^{r} x_{ii} - \sum_{i=1}^{r}(x_{i+}x_{+i})}{N^2 - \sum_{i=1}^{r}(x_{i+}x_{+i})}$$

where r is the number of rows in the accuracy matrix, x_{ii} is the number of observations in row i and column i, x_{i+} and x_{+i} are the marginal totals of row i and column i, respectively, and N is the total number of observations. This is a more complete way of expressing the overall accuracy in terms of a single number, as it accounts for both the omission and commission errors in a balanced way. The number labeled "Kappa variance" is an estimate of the variance of the estimated Kappa statistic.

25 Cohen, J. A coefficient of agreement form nominal scales, *Education Psychological Measurement*, Vol. 20, No. 1, pp. 37-46, 1960,.
26 Congalton, Russell G., A Review of Assessing the Accuracy of Classifications of Remotely Sensed Data, *Remote Sensing of Environment*, Vol. 37, pp. 35-46, 1991.

3.17 An Example Analysis

We illustrate the analysis of a multispectral data set using algorithms discussed so far. The data set used is of moderate dimensionality. The illustration is conducted using the MultiSpec© system.[27] For brevity and clarity, not all of the steps mentioned above are shown for this example.

<u>The Data Set</u>

Flightline C1 (FLC1, see Figure 3-42), a historically significant data set, is over an agricultural area located in the southern part of Tippecanoe County, Indiana. Though collected with an airborne scanner in June

Band #	Wavelength, μm
1	0.40 - 0.44
2	0.44 - 0.46
3	0.46 - 0.48
4	0.48 - 0.50
5	0.50 - 0.52
6	0.52 - 0.55
7	0.55 - 0.58
8	0.58 - 0.62
9	0.62 - 0.66
10	0.66 - 0.72
11	0.72 - 0.80
12	0.80 - 1.00

Table 3-7. Spectral Bands for FL:C1.

Figure 3-42. FLC1. (In color on CD)

---

[27] MultiSpec for Macintosh and Windows computers is available along with substantial documentation on its use at *http://dynamo.ecn.purdue.edu/ ~biehl/MultiSpec/.*

1966, this data remains contemporary. Key attributes that make it valuable, especially for illustrative purposes, are that it has more than a few spectral bands (12 bands), contains a

Classes
background
Corn
Soybeans
Alfalfa
Bare Soil
Farmstead
Oats
Pasture
Rye
Red Clover
Hay (Timothy)
Wheat

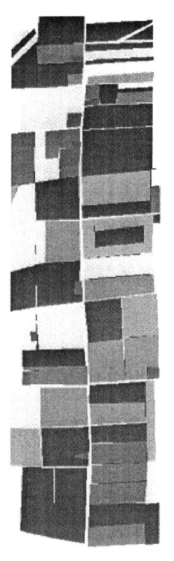

significant number of vegetative species or ground cover classes (at least 9), includes many regions (e.g., fields) containing a large numbers of contiguous pixels from a given class (thus facilitating quantitative results evaluation), and has "ground truth" available. The spectral bands in the data set are given in Table 3-7

The data set consists of 949 scan lines with 220 pixels per scan line, or 208,780 pixels. The scanner used had an instantaneous field of view (IFOV) of 3 milliradians and was flown at an altitude of approximately 2600 ft above terrain. The sensor scans approximately ± 45° about nadir with somewhat less than that being digitized. Each pixel was digitized to 8-bit precision. No radiometric adjustments have been made to

Fig3-43. Ground truth map for Flightline C1.(In color on CD)

the data for atmospheric or other observational effects. The data geometry has also not been adjusted; since the analysis classifies the pixels individually, distortions in geometry have no effect on the accuracy of classification.

Figure 3-43 provides "ground truth" for the flightline in the form of a designation of thecrop species or ground cover type for each area of the flightline. The information was collected at the time the overflight took place.

	Field Name	Class	First Line	Last Line	First Col.	Last Col.	No. of Samples
1	Alfalfa1	1	731	737	129	177	343
2	Alfalfa2	1	749	755	131	171	287
3	Alfalfa3	1	809	817	155	183	261
4	Soil1	2	97	119	49	85	851
5	Corn1	3	167	177	33	77	495
6	Corn2	3	267	283	45	61	289
7	Corn3	3	319	341	21	31	253
8	Corn4	3	603	625	13	33	483
9	Oats1	4	421	455	63	83	735
10	Oats2	4	591	599	135	181	423
11	RedCl1	5	439	447	139	183	405
12	RedCl2	5	539	565	175	195	567
13	RedCl3	5	599	619	69	95	567
14	Rye1	6	527	569	127	155	1247
15	Soy1	7	65	81	69	89	357
16	Soy2	7	237	253	141	167	459
17	Soy3	7	307	327	59	81	483
18	Soy4	7	773	777	135	179	225
19	Water	8	3	4			43
			1	1			
			2	13			
			9	1			
20	Wheat1	9	295	303	134	175	378
21	Wheat2	9	471	495	172	201	750
22	Wheat3	9	607	665	203	211	531
23	Wheat-2-1	10	655	695	17	41	1025
	Total Number of training samples						11,457

Figure 3-8 Training samples for the analysis

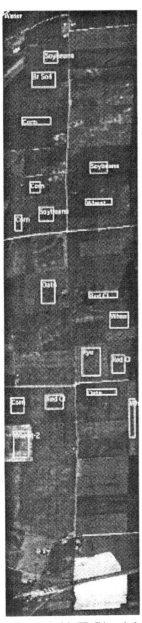

Figure 3-44. FLC1 training areas. (In color on CD)

Procedure

1. The most important step in the analysis of any data set is specifying the classes that are to be used. This must be done quantitatively and is usually done by marking examples of each class in the data to be analyzed. The statistical model for each class can be calculated from these samples. Samples making up the training set for this example are listed in Table 3-8 and marked in Figure 3-44.

   Note that in this case two subclasses were defined for the class Wheat, but all other classes were modeled with a single Gaussian density function.

1	10
2	1,9
3	1,6,9
4	1,6,9,10
5	1,6,9,10,12
6	1,2,6,9,10,12
7	1,2,6,8,9,10,12
8	1,2,6,8,9,10,11,12
9	1,2,3,6,8,9,10,11,12
10	1,2,3,4,6,8,9,10,11,12
11	1,2,3,4,5,6,8,9,10,11,12
12	All 12

Table 3-9. Spectral band subsets for the analysis.

2. The Feature Selection processor was run for feature subsets of size $1 \leq N \leq 12$. The feature subsets were ordered according to the largest minimum Bhattacharyya distance. The first-listed feature subset was chosen in each case, given in Table 3-9.

3. The flightline was classified using the Quadratic Maximum Likelihood algorithm with each of the feature subsets. The training and test sample accuracy was determined using the test set listed in Table 3-10. This test set was selected to be as comprehensive as possible but constrained to areas that were reasonably free from anomalous pixels such as bare areas, and weeds.

Table 3-11 provides a listing of the performance for the test samples. In this case the two subclasses of wheat have been grouped into one class labeled Wheat. It is seen from this table that all classes were fairly accurately classified except there was significant confusion between Alfalfa and Red Clover.

Field Number	First Line	Last Line	First Column	Last Column	Class	No. of Samples
1	57	89	47	103	Soybeans	1881
2	63	74	115	169	Soybeans	660
3	93	101	113	183	Soybeans	639
4	123	133	43	101	Soybeans	649
5	133	149	43	83	Soybeans	697
6	217	273	109	201	Soybeans	5301
7	705	797	69	111	Soybeans	3999
8	291	341	43	92	Soybeans	2550
9	489	519	115	161	Soybeans	1457
10	643	663	125	197	Soybeans	1533
11	647	659	51	87	Soybeans	481
12	647	675	93	111	Soybeans	551
13	705	797	33	61	Soybeans	2697
14	759	785	121	197	Soybeans	2079
15	157	187	17	101	Corn	2635

Table 3-10. Test Set for FLC1 begun.

Field First Number Line		Last Line	First Column	Last Column	Class	No. of Samples
16	189	215	17	79	Corn	1701
17	221	255	39	55	Corn	595
18	261	287	39	65	Corn	729
19	307	349	14	35	Corn	946
20	401	421	111	194	Corn	1764
21	589	643	3	43	Corn	2255
22	327	335	109	197	Oats	801
23	365	377	131	185	Oats	715
24	413	467	45	91	Oats	2585
25	583	605	121	191	Oats	1633
26	285	317	109	199	Wheat	3003
27	347	353	107	205	Wheat	693
28	385	393	109	203	Wheat	855
29	459	509	167	211	Wheat	2295
30	581	689	203	211	Wheat	981
31	649	699	3	43	Wheat	2091
32	129	133	113	199	Red Clover	435
33	357	399	61	95	Red Clover	1505
34	433	453	113	197	Red Clover	1785
35	521	561	173	215	Red Clover	1763
36	559	581	49	109	Red Clover	1403
37	589	633	49	109	Red Clover	2745
38	613	619	121	183	Red Clover	441
39	629	637	123	191	Red Clover	621
40	675	695	127	195	Red Clover	1449
41	729	737	121	195	Alfalfa	675
42	745	757	121	195	Alfalfa	975
43	793	815	121	195	Alfalfa	1725
44	525	577	119	163	Rye	2385
45	137	149	87	101	Bare Soil	195
46	95	117	45	89	Bare Soil	1035
					Total	70,588

Table 3-10. Test Set for FLC1 continued.

| Project Group Name | Accur.+ (%) | No. Samp. | Number of Samples in Thematic Image Group ||||||||||
			1 bkgnd	2 Alfalfa	3 Br Soil	4 Corn	5 Oats	6 Red Cl	7 Rye	8 Soybeans	9 Water	10 Wheat
Alfalfa	89.6	3375	0	3024	0	22	144	180	0	5	0	0
Br Soil	99.3	1230	0	0	1221	1	0	0	0	8	0	0
Corn	92.1	10625	0	16	0	9786	123	275	1	423	0	1
Oats	88.4	5781	0	74	0	21	5112	244	1	24	0	305
Red Cl	85.6	12147	0	1295	0	223	151	10393	1	76	0	8
Rye	98.2	2385	0	0	0	0	8	0	2341	1	0	35
Soybeans	97	25174	0	31	5	434	158	10	46	24429	0	61
Water	Results are not available.											
Wheat	99.5	9918	0	0	0	0	16	0	34	0	0	9868
TOTAL		70635	0	4440	1226	10487	5712	11102	2424	24966	0	10278
Reliability Accuracy (%) **				68.1	99.6	93.3	89.5	93.6	96.6	97.8	0	96

OVERALL GROUP PERFORMANCE (66174 / 70635) = 93.7%
Kappa Statistic (X100) = 92.0%. Kappa Variance = 0.000001.
+ (100 - percent omission error); also called producer's accuracy.
* (100 – percent commission error); also called user's accuracy.

Table 3-11. Classifier performance as measured by the test samples.

The results of this Quadratic Maximum Likelihood pixel classification in terms of the training and test set accuracies for the various subset of Table 3-9 are shown in Figure 3-46. Note that the training set contains 11,457 samples with no class smaller than 851 samples. The one exception to this is the class Water, which is very small and very spectrally distinct. This training set size is very much larger than would ordinarily be available to the analyst. Thus no Hughes effect is evident. Note also that the test set contains 70,588 samples, almost seven times the size of the training set and nearly one-third of the total of 208,780 pixels of the entire flightline.

188

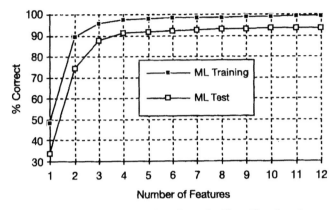

Figure 3-46. FLC1 Training and Test Sample Classification Accuracy

Figure 3-47 shows a thematic map presentation of the results. Given the quantitative evaluation above, from a subjective standpoint the results shown in this figure appear to be reasonably good, although there are some random appearing "salt and pepper" errors. A comparative look at this thematic map presentation of results with the image space view of the original data shows that some of the "errors" may not even be errors. Areas such as sod waterways in fields, weed areas, low spots with poor drainage, and other such anomalous occurrences result in variations in the type of ground cover in otherwise "uniform" fields. Furthermore, variations show up because of soil conditions under the crops, resulting in variations in the quality of the canopy of the crop. From this, it becomes clear that simply calling a few pixels to be used for training as corn, for example, may not fully specify what the analyst wishes to include in the class Corn.

Figure 3-48 gives a "probability map" or "likelihood map" for the classification shown in Figure 3-47b. In this presentation, color is used to indicate the level of likelihood that a given pixel assigned to a certain class is actually a member of that class. Darker blue

189

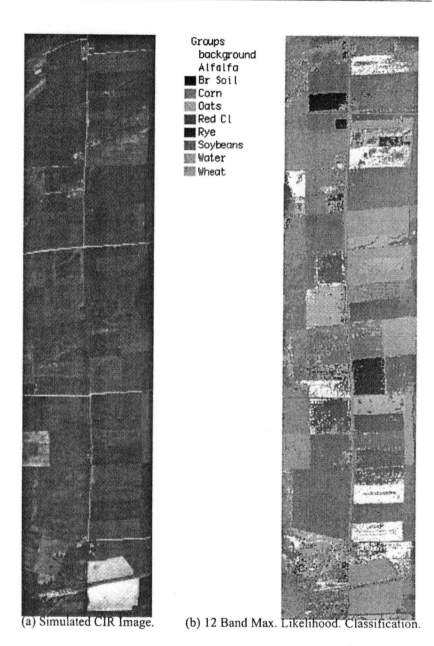

(a) Simulated CIR Image.   (b) 12 Band Max. Likelihood. Classification.

Figure 3-47. A comparison of the classification result in thematic map form
with the original data in image space. (In color on CD)

Figure 3-48. Probability map or likelihood map of the classification shown in Figure 3-47B. (In color on CD)

indicates the lowest likelihood, and the range extends up through yellow and finally to red for the highest likelihood. This is useful, for example, to determine if any significant classes of ground cover have been overlooked in establishing a presumably exhaustive list of classes. In this example, classes were not established for the road, which runs down the center of the flightline and several crossroads, and for farmsteads. The figure also shows that there is a large field at the bottom of the flightline beyond the area for which ground truth data was collected, that is not a strong member of the class Wheat to which it has been assigned, even though, from Figure 3-47b, its pixels have been quite uniformly assigned to the class Wheat. Attention is also drawn to an irregularly shaped area about halfway down the flightline and at the extreme right edge, which is shown in medium blue. This is a cloud shadow area, and the shadowing is evidently the cause of a portion of this area to be assigned to the wrong class. From these observations it is perhaps apparent that these likelihood maps, or probability maps, can be quite useful in assessing the quality of a given classification. Had a threshold been used with the classification, it is these dark blue areas that would have been thresholded.

Concluding Remarks

This example analysis is intended to point to some of the possible processing steps in analysis of a multispectral data set and illustrate their likely impact in a typical situation. Further, enough information is provided so that the reader, using the copy of MultiSpec, its documentation, and the FLC1 data set available on the acompanying CD, could reproduce the results obtained and try other options.

It is clear from such exercises that, no matter what algorithms are used to analyze multispectral data, the aspect of greatest importance is the accurate and thorough quantitative modeling of the user's classes of interest, relative to the other spectral responses that exist in the data set, and doing so in such a manner as to maximize the separability between them in feature space. This fact places great emphasis upon the analyst and his/her skill and knowledge about the scene. The current goal of the on-going research effort in this field thus is to find ever-improving tools to assist in this process and increase its objectivity.

Part III Additional Details

Although this book is printed in black and white, figures where color is necessary or helpful are available for view in color on the accompanying compact disk (CD). The color figures that are available on the CD are noted in the figure captions accordingly.

4 Training a Classifier

In this chapter further details about the training of a classifier are presented. The basic requirements for an effective training set are discussed, as well as ways for making a training set more robust. Algorithms are presented that improve the ability of a training set to generalize and for dealing with situations where only a very small set of training samples are available.

4.1 Classifier Training Fundamentals

As was seen in Chapter 3, the most critical factor in obtaining a satisfactory result is the establishment of proper classifier training, one that provides an adequately precise quantitative description of the information desired from the data set. There are two basic parts to this training process, the selection of an appropriate class list and an adequately precise mathematical description of each class. The class list must be based on both the data set in hand and the information desired from it. The adequacy of the mathematical class descriptions is usually a function of the number and typicality of the training samples. Fundamentally, for a classifier to be well trained, the classes in the class list must be

- **Of informational value**. The list must contain at least the classes of interest to the user.
- **Separable** (using the available features). No matter how badly a class may be desired, there is no point in including it in the class list if it is beyond the discriminating power of the available feature set.
- **Exhaustive**. There must be a logical class to which to assign every pixel in the scene.

An equivalent statement to this is that a well trained classifier must have successfully modeled the distribution of the <u>entire</u> data set, but it must done in such a way that the different classes of interest to the user are as distinct from one another as possible. What is desired in mathematical terms is to have the density function of the entire data set modeled as a mixture of class densities, namely

$$p(\mathbf{x} \mid \theta) = \sum_{i=1}^{m} \alpha_i p_i(\mathbf{x} \mid \phi_i) \qquad (4\text{-}1)$$

where $\mathbf{x}$ is the measured feature (vector) value, p is the probability density function describing the entire data set to be analyzed, θ symbolically represents the parameters of this probability density function, p_i is the density function of class i desired by the user with its parameters being represented by ϕ_i, α_i is the weighting coefficient or probability of class i, and m is the number of classes. The parameter sets θ and ϕ_i are to be discussed next in the context of what limitations are appropriate to the form of these densities.

As has been seen, the Gaussian model for class distributions is very convenient, as the Gaussian density function has many convenient properties and characteristics, both theoretically and practically. Chief among these convenient properties is that its use requires knowledge of only the first two statistical moments, the mean vector and the covariance matrix, both of which can usually be estimated adequately from a reasonable sized training set. However, one cannot always assume that classes of interest will be Gaussianly distributed, and a more flexible and general model is needed. On the other hand, a purely nonparametric approach is frequently not tractable in the remote sensing situation, as very large training sets are usually required to provide precise enough estimates of nonparametric class densities, while, it is a characteristic of the remote sensing application of pattern recognition theory that there will be a paucity of training samples.

As has been seen, a very practical approach is to model each class density by a linear combination of Gaussian densities. Thus, rather than having a single component of equation (4-1) represent a single user class, the number of components in the combination can be increased, with different subsets of them representing each user class. In this way a very general capability can be provided, while still maintaining the convenient properties of the Gaussian assumption. Indeed, theoretically every smooth density function can be approximated to within any accuracy by such a mixture of Gaussian densities.

Thus for a well-trained classifier, $p(\mathbf{x} \mid \theta)$, the probability density function of the entire data set, can be modeled by a combination of m Gaussian densities. Assume that there are J (user) classes in the feature space denoted by $S_1...,S_J$, $J \leq m$. Each class consists of a different subset of the m Gaussian components, and indicate that component i belongs to class S_j by $i \in S_j$. Thus in equation (4-1), $\phi_i = (\mu_i, \Sigma_i)$, the mean vector and covariance matrix, respectively, of Gaussian component i, and $\theta = (\alpha_1, ... , \alpha_m, \mu_1, ... , \mu_m, \Sigma_1 ... , \Sigma_m)$. All of the conditions required above are met if θ can be determined so that equation (4-1) is satisfied in such a way that the J classes are adequately separable and correspond to user classes. Determining these m components is what must be accomplished by the training phase.

How can this be accomplished in practice? Return to the three fundamentals stated above.

- The first of the three–classes must be of informational value–is serviced by the user defining the classes through their training samples accordingly.

- The second one–classes must be separable–is supported by the use of clustering algorithms, distance measures and other separability metrics in conjunction with the training process.

- The third can be assisted, for example, by the use of a likelihood map from a preliminary classification–a map showing, on a relative scale, how likely each pixel is to actually be a member of the class to which it has been assigned. This allows one to locate pixels in the data set that, while having been assigned to a class, the strength of the class association is very limited, as evidenced by the likelihood of the most likely class being very low.

However, to this point, none of these tools guarantees that equation (4-1) is indeed satisfied, that the combination of class conditional density functions defined in this process provide a good model of the entire data set to be analyzed.

4.2 The Statistics Enhancement Concept

It is this problem, the degree to which the left side of equation (4-1) equals the right side, that the so-called Statistics Enhancement process is designed to solve. Briefly, the idea is to carry out an iterative calculation based on the training samples together with a systematic sampling of all the other (unlabeled) pixels in the scene that will adjust or "enhance" the statistics so that, while still being defined by the training samples, the collection of class conditional statistics better fit the entire data set.[28–32] This amounts to a hybrid, supervised/unsupervised training scheme.

From the application point of view, this process can have several possible benefits:

1. The process tends to make the training set more robust, providing an improved fit to the entire data set and thus providing improved generalization to data other than the training samples.

2. The process tends to mitigate the Hughes phenomenon, namely the effect that for a finite sized training set, as one adds more features, the accuracy increases for a while, then peaks and begins to decrease. This problem of declining accuracy for increasing numbers of features is especially of concern when either (a) the training set size is small, or (b) the number of features available is large or both. Enhancing the statistics by this scheme tends to move the peak accuracy vs. number of features to a higher

28 B. M. Shahshahani and D. A. Landgrebe, Using partially labeled data for normal mixture identification with application to class definition, *Proceedings of the International Geoscience and Remote Sensing Symposium (IGARSS'92)*, Houston, TX, pp. 1603–1605 May 26-29, 1992.

29 B.M. Shahshahani, D.A. Landgrebe, On the Asymptotic Improvement of Supervised Learning by Utilizing Additional Unlabeled Samples; Normal Mixture Density Case, *SPIE Int. Conf. Neural and Stochastic Methods in Image and Signal Processing*, San Diego, CA, July 19-24, 1992.

30 B. M. Shahshahani and D. A. Landgrebe, Use of unlabeled samples for mitigating the Hughes phenomenon, *Proceedings of the International Geoscience and Remote Sensing Symposium (IGARSS'93)*, Tokyo, pp. 1535-1537, August 1993.

31 B. M. Shahshahani and D. A. Landgrebe, Classification of Multi-Spectral Data By Joint Supervised-Unsupervised Learning, PhD thesis, School of Electrical Engineering, Purdue University, December 1993, School of Electrical Engineering Technical Report TR-EE-94-1, January 1994.

32 B. M. Shahshahani and D. A. Landgrebe, The Effect of Unlabeled Samples in Reducing the Small Sample Size Problem and Mitigating the Hughes Phenomenon, *IEEE Transactions on Geoscience and Remote Sensing*, Vol. 32, No. 5, pp. 1087-1095, September 1994.

value at a higher dimensionality, thus allowing one to obtain greater accuracy with a limited training set.

3. An estimate is obtained for the prior probabilities of the classes, as a result of the use of the unlabeled samples, something that cannot be done with the training samples alone.

4.3 The Statistics Enhancement Implementation

To carry out the process, for each class S_j, assume there are N_j training samples available. Denote these samples by z_{jk} where $j=1,...,J$ indicates the class of origin and $k = 1,...,N_j$ is the index of each particular sample. The training samples are assumed to come from a particular class without any reference to the exact component within that class. In addition to the training samples, assume that N unlabeled samples, denoted by x_k, $k=1,...,N$, are also available from the mixture.

The process to be followed is referred to as the EM[33, 34] (expectation maximization) algorithm. The procedure is to maximize the log likelihood to obtain ML estimates of the parameters involved. The log likelihood expression to be maximized can be written in the following form:[35]

---

33 A.P. Dempster, N.M. Laird, D.B. Rubin, Maximum Likelihood Estimation from Incomplete Data via EM Algorithm, *Journal of the Royal Statistical Society*, B 39, pp. 1-38, 1977.

34 R.A. Redner, H.F. Walker, Mixture Densities, Maximum Likelihood and the EM Algorithm, *SIAM Review*, Vol. 26, No. 2, pp. 195-239, 1984.

35 The form implemented in MultiSpec actually assumes one component per class. Thus each subclass is treated as if it were a class.

$$L(\theta) = \sum_{k=1}^{N} \log p(x_k \mid \theta) + \sum_{j=1}^{J} \sum_{k=1}^{N_j} \log \left(\frac{1}{\sum_{t \in S_j} \alpha_t} \sum_{l \in S_j} \alpha_l p_l(z_{jk} \mid \phi_l) \right) \tag{4-2}$$

where x_k is the kth unlabeled sample, and z_{jk} is the kth training sample of class j. The first term in this function is the log likelihood of the unlabeled samples with respect to the mixture density. The second term indicates the log likelihood of the training samples with respect to their corresponding classes of origin. The EM equations for obtaining the ML estimates are the following:[36]

$$\alpha_i^{+} = \frac{\displaystyle\sum_{k=1}^{N} P^c(i \mid x_k) + \sum_{k=1}^{N_j} P_j^c(i \mid z_{jk})}{N \left(1 + \dfrac{N_j}{\displaystyle\sum_{r \in S_j} \sum_{k=1}^{N} P^c(r \mid x_k)} \right)} \tag{4-3}$$

$$\mu_i^{+} = \frac{\displaystyle\sum_{k=1}^{N} P^c(i \mid x_k) x_k + \sum_{k=1}^{N_j} P_j^c(i \mid z_{jk}) z_{jk}}{\displaystyle\sum_{k=1}^{N} P^c(i \mid x_k) + \sum_{k=1}^{N_j} P_j^c(i \mid z_{jk})} \tag{4-4}$$

---

36 For details, see B. M. Shahshahani and D. A. Landgrebe, Classification of Multi-Spectral Data By Joint Supervised-Unsupervised Learning, PhD thesis, School of Electrical Engineering, Purdue University, December 1993, School of Electrical Engineering Technical Report TR-EE-94-1, January 1994.

$$\Sigma_i^+ = \frac{\displaystyle\sum_{k=1}^{N} P^c(i \mid x_k)(x_k - \mu_i^+)(x_k - \mu_i^+)^T + \sum_{k=1}^{N_j} P_j^c(i \mid z_{jk})(z_{jk} - \mu_i^+)(z_{jk} - \mu_i^+)^T}{\displaystyle\sum_{k=1}^{N} P^c(i \mid x_k) + \sum_{k=1}^{N_j} P_j^c(i \mid z_{jk})}$$

$$(4\text{-}5)$$

The equations are applied iteratively with respect to the training and unlabeled samples where "c" and "+" refer to the current and next values of the respective parameters, $i \in S_j$, and $P^c(.|.)$ and $P^c_j(.|.)$ are the current values of the posterior probabilities:

$$P^c(i \mid x_k) = \frac{\alpha_i^c f_i(x_k \mid \mu_i^c, \Sigma_i^c)}{f(x_k \mid \theta^c)} \qquad (4\text{-}6a)$$

$$P_j^c(i \mid z_{jk}) = \frac{\alpha_i^c f_i(z_{jk} \mid \mu_i^c, \Sigma_i^c)}{\displaystyle\sum_{i \in j} \alpha_i^c f_i(z_{jk} \mid \mu_i^c, \Sigma_i^c)} \qquad (4\text{-}6b)$$

Thus, as the iteration proceeds, successively revised values for the mean, covariance, and weighting coefficient of each component of each class are arrived at which steadily approach the values for a maximum of the expected likelihood value for the mixture density.

4.4 Illustrations Of The Effect Of Statistics Enhancement

To illustrate the effect of the statistics enhancement process empirically, consider the FLC 1 12-band data set for which 10 classes have been defined with training samples. It is not possible to visualize the data and its training in 12-dimensional space, but we can see the effect of the enhancement process even in a single band. Shown in Figure 4-1 is a histogram of the data in band 12. The original data is at 8-bit precision and thus the abscissa is labeled

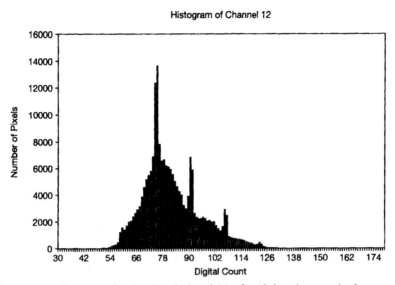

Figure 4-1. Histogram for the data in band 12 of a 12-band example data set.

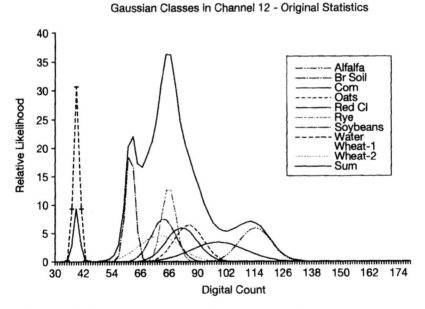

Figure 4-2. Histograms of the individual classes for the data of an example data set. (In color on CD)

203

"digital count" numbers on a scale of 0 to 255. Only a part of this scale is needed for the band 12 histogram as shown.

During the training process, 10 classes and subclasses were defined, and their mean vectors and covariance matrices were calculated for each. Figure 4-2 shows a Gaussian density for each of the 10 classes in band 12, using its calculated mean and variance. This graph also shows a curve labeled "Sum," which is a linear combination of the individual Gaussian curves like equation (4-1) above. However, since the probabilities of each class cannot be known at this point in the training process, the weighting coefficients used to form the Sum graph are proportional to the number of training samples used for each class. It is seen that this Sum graph does not look much like the histogram of Figure 4-1, implying that at this point, the right side of equation (4-1) does not equal the left side.

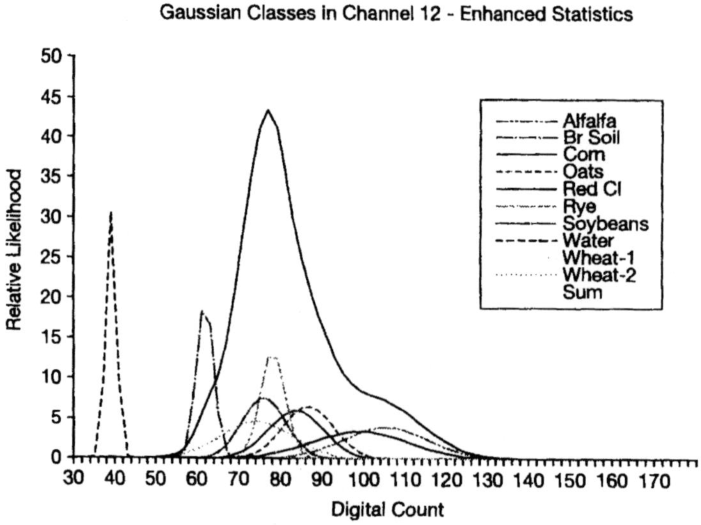

Figure 4-3. Densitygraphs for the enhanced class statistics of the example. (In color on CD)

204

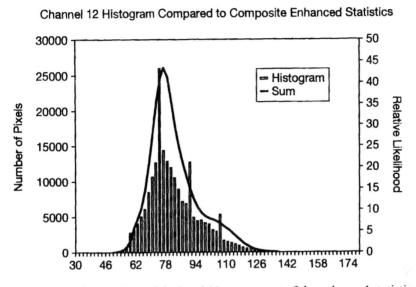

Figure 4-4. Comparison of the band 12 component of the enhanced statistics with the original histogram

Applying the Statistics Enhancement process to the class statistics results in the individual class statistics being modified to some extent, as seen by comparing Figure 4-2 above with that showing the enhanced statistics of Figure 4-3. Notice that the mean value and variance of some of the individual classes have been changed perceptibly. One also obtains an estimate of the proportions (or probabilities) of each class, which are the α_i coefficients of equation (4-1). These were used to calculate the new Sum curve of the graph of Figure 4-3.

And finally, Figure 4-4 compares the new Sum curve to the histogram of the data. It is seen that these two compare much more favorably as a result of the Statistics Enhancement process.

An illustration of additional characteristics of the use of Statistics Enhancement is shown in Figure 4-5. Here a 12 class, 7 band classification was carried out on a small Thematic Mapper data

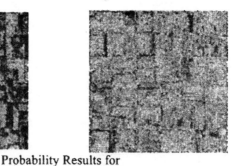

Probability Results for
Original Statistics   Enhanced Statistics

Figure 4-5. An example showing one of the effects of Statistics
   Enhancement. The relative likelihood of class membership for
   each pixel after classification is color coded from dark blue (low
   likelihood) through yellow to orange (high likelihood). (In color
   on CD)

set. The classification map of the results showed good accuracy
when compared to the ground truth information. Figure 4-5 gives a
comparison of the likelihood maps before and after the Enhance
Statistics step. A substantial improvement is noted in the degree of
membership indicated for each pixel, indicating the collection of
class statistics now fit the entire data set much more closely.
Quantitatively, after classification, the average likelihood value of
28.5% using the original statistics was increased to 49.1% by the
enhancement process.

To illustrate the effect of this procedure on the Hughes effect, the
results of two experiments will be presented. The first was conducted
with a portion of an AVIRIS data set consisting of 210 bands taken
over Tippecanoe County, Indiana. Four ground cover classes were
determined by consulting the ground truth map. The classes were

bare soil (380 pixels), wheat (513 pixels), soybean (741 pixels), and corn (836 pixels), using 20 randomly drawn training samples. Every fifth band was used and the bands were ranked by sequential use of average Bhattacharyya distance. An additional number of unlabeled samples were used via the equations above for estimating the parameters. Subsequently the rest of the samples were classified according to the (Bayes) MAP decision rule. The experiment was performed once with 500 unlabeled samples and once with 1000 unlabeled samples.

The second experiment was conducted on a portion of the Flight Line C1 (FLC1) data set, which is the 12 band airborne multispectral data set of the example at the end of Chapter 3. Four ground-cover classes were determined using the ground truth map: corn (2436 pixels), soybean (2640 pixels), wheat (2365 pixels) and red clover (2793 pixels). Bands were again ranked by sequentially applied average pair wise Bhattacharya distance. The experiment was conducted using 15 randomly drawn training samples. A number of unlabeled samples were used simultaneously for estimating the parameters, and again an MAP classification of the remaining samples was performed subsequently. Experiment 2 was performed once with 100 and once with 500 unlabeled samples. Note that as previously pointed out, an additional benefit of using unlabeled samples is that the prior probabilities of the classes can be obtained. Therefore instead of the ML classifier, the MAP classifier can be constructed. Without the unlabeled samples, generally the prior probabilities cannot be estimated because the training samples are usually obtained separately from each class.

Each experiment was repeated 10 times, and Figures 4-6 and 4-7 show the results. In both of these figures, the curves for the case where only training samples alone are used also shown for comparison and are labeled "supervised."[37]

From Figures 4-6 and 4-7 it can be seen that the use of additional unlabeled samples in the learning process can enhance the classification performance when the dimensionality of data begins to approach the number of training samples. In Figure 4-6, the Hughes phenomenon performance decrease that began around dimension 8 when supervised learning is used, is delayed to dimension 16 when 500 or 1000 additional unlabeled samples are incorporated. Meanwhile, the minimum error for the supervised

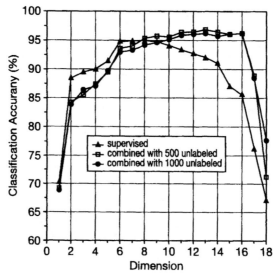

Figure 4-6. Effect of additional unlabeled samples on the classification performance for experiment 1 (AVIRIS data) with 20 training samples/class.

---

[37] The graphs published in B. M. Shahshahani and D. A. Landgrebe, Use of unlabeled samples for mitigating the Hughes phenomenon, *Proceedings of the International Geoscience and Remote Sensing Symposium (IGARSS'93)*, Tokyo, pp. 1535-1537, August 1993 are corrected here. In the former paper, the tails of the curves were shown incorrectly to decay too rapidly, hence the Hughes phenomenon was exaggerated.

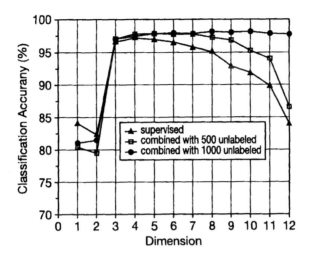

Figure 4-7. Effect of additional unlabeled samples on the classification performance for experiment 2 (FLC1 data) with 15 training samples/ class.

learning case was 5.42% and was achieved at dimension 7. For the cases with an additional 500 and 1000 unlabeled samples, the minimum errors were 3.11% and 3.78% at dimensions 13, and 16 respectively. The use of additional unlabeled samples not only delayed the occurrence of the Hughes phenomenon but also made the information in the new features usable for decreasing the error further. Similarly, in Figure 4-7, the Hughes phenomenon that began at around dimension 4 was delayed to around dimension 7 by using 100 unlabeled samples and was virtually eliminated by using 500 additional unlabeled samples. The minimum error for the supervised learning case was 2.77% at dimension 4. For the cases with additional 100 and 500 unlabeled samples the minimum errors were 2.20% and 1.88% at dimensions 5 and 10 respectively.

4.5 Robust Statistics Enhancement[38, 39]

When the ratio of the number of training samples to the dimensionality is small, parameter estimates become highly variable, causing the deterioration of classification performance. This problem has become more prevalent in remote sensing with the emergence of newer generations of sensors with as many as several hundred spectral bands. While the new sensor technology provides higher spectral detail, enabling a greater number of spectrally separable classes to be identified, the needed labeled samples for designing the classifier are sometimes difficult and expensive to acquire. As was seen above, better parameter estimates can be obtained by exploiting a large number of unlabeled samples in addition to training samples using the expectation maximization algorithm under the mixture model.

However, the estimation method is sensitive to the presence of statistical outliers, namely samples that are spectrally significantly different than any of the defined classes. In remote sensing data, miscellaneous classes with few samples are common, are often difficult to identify, and may constitute such statistical outliers. The existence of such samples can substantially negatively impact the statistics enhancement process. Therefore it may be desirable to use a robust parameter estimation method for the mixture model. The method assigns full weight to training samples but automatically gives reduced weight to unlabeled samples.

---

[38] S. Tadjudin and D. A. Landgrebe, Robust parameter estimation for mixture model, *IEEE Transactions on Geoscience and Remote Sensing*, Vol. 38, No. 1, pp. 439-445, January 2000.

[39] S. Tadjudin, Classification of high dimensional data with limited training samples, PhD thesis, School of Electrical and Computer Engineering, Purdue University, 1998.

The basic EM algorithm first estimates the posterior probabilities of each sample belonging to each of the component distributions, and then computes the parameter estimates using these posterior probabilities as weights. With this approach each sample is assumed to come from one of the component distributions, even though it may greatly differ from all components. The robust estimation attempts to circumvent this problem by including the typicality of a sample with respect to the component densities in updating the estimates in the EM algorithm. A series of illustrations will demonstrate how this robust method prevents performance deterioration due to statistical outliers in the data as compared to the estimates obtained from the original EM approach described above.

4.6 Illustrative Examples of Robust Expectation Maximation

In the following examples, the performances of quadratic classifiers are compared: parameters estimated from training samples alone (ML), the original EM algorithm (EM), and the robust form of the algorithm (REM).

Experiments 1 through 4 are performed using a portion of an AVIRIS data set taken over NW Indiana's Indian Pine test site in the month of June. The scene contains four information classes: corn-no till, soybean-no till, soybean-min till, and grass. By visual inspection of the data in image space, the list of these ground cover types is assumed to be exhaustive. A total of 20 channels from the water absorption region and noisy bands (104-108, 150-163, 220) are removed from the original 220 spectral channels, leaving 200 spectral features for the experiments. An image of the test data and the ground truth map are shown in Figure 4-8. The number of labeled samples in each class is shown in Table 4-1. The numbers of spectral channels were selected to be 10, 20, 50, 67, and 100. These channels were selected by sampling the spectral range at fixed intervals.

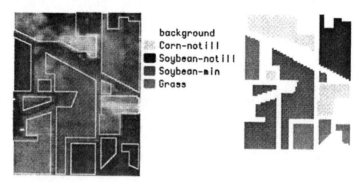

Figure 4-8. A portion of AVIRIS Data and Ground Truth Map.
(In Color on CD).

Training samples were randomly selected, and the remaining labeled samples were used for testing. The algorithms are repeated for 10 iterations and the classification is performed using the quadratic maximum likelihood classifier. The maximum likelihood (ML) method using only the training samples to estimate the parameters is denoted as ML in the following experiments.

Class Names	No. of Labeled Samples
Corn-no till	910
Soybean-no till	638
Soybean-min till	1421
Grass	618

Table 4-1. Class Description for AVIRIS Data.

Illustration 1

This first illustration is intended to compare EM and REM without outliers in the data. To obtain data without outliers, synthetic data is generated using the statistics computed from the labeled samples of the four classes. A total of 2000 test samples per class is generated,

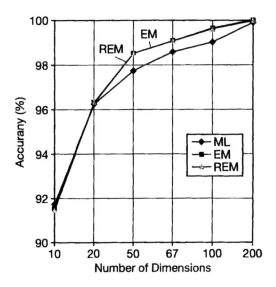

Figure 4-9. Mean Accuracy for Experiment 1 with 500 Training Samples and 1500 Test Samples.

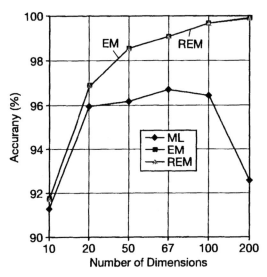

Figure. 4-10. Mean Accuracy for Experiment 2 with 250 Training Samples and 1500 Test Samples.

500 of which are used as the training samples. Since the training samples are selected at random, the experiment is repeated 5 times and the mean classification accuracy is recorded. The mean accuracy is shown in Figure 4-9.

The results show that when no outliers are present in the data, the EM and REM algorithms have similar performance and both result in better performance than the maximum likelihood classifier using the training samples alone. Since there are many design samples available, the best performance is obtained at 200 features.

Illustration 2

The synthetic data from Illustration 1 are used but only 250 training samples are selected for each class. The number of test samples is kept at 1500. Again, no outliers are present in the data. The results are shown in Figure. 4-10.

With fewer training samples, the performance of the maximum likelihood classifier (ML) using the training samples deteriorates. The decline is particularly obvious at higher dimensionality. Compared to the previous experiment, the accuracy has dropped 7% at 200 features. However, when unlabeled samples are used for the mixture model, the performance remains stable even when the number of training samples declines. The results again show that when no outliers are present in the data, the EM and REM algorithms have comparable performance and both achieve better classification accuracy than the ML classifier without using additional unlabeled samples.

Illustration 3

The peviolus illustration 2 is repeated with only 400 test samples generated for each class. The number of training samples per class is 250. Again, no outliers are present in the data. The results are

214

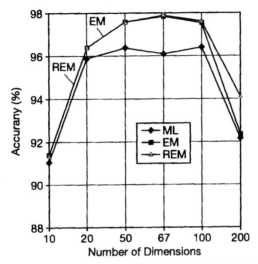

Figure. 4-11. Mean Accuracy for Experiment 3 with 250 Training Samples and 400 Test Samples.

shown in Figure 4-11. Compared to the results from two previous illustrations in which many more unlabeled samples were used, the classification results for all three methods deteriorate in this illustration. This deterioration is manifested as the Hughes phenomenon. Hence, the likelihood parameter estimation for the mixture model is shown to be affected by the number of unlabeled samples relative to dimensionality. Specifically, it implies that 650 samples are still inadequate to characterize these 200-dimensional Gaussian distributions. The results again indicate that without outliers, the EM and REM algorithms have comparable performance and both have better classification accuracy than the ML classifier without using additional unlabeled samples.

Illustration 4

This illustration is conducted using the real samples from the data. Again, since all four classes are represented by the training samples,

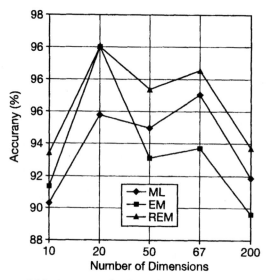

Figure. 4-12. Accuracy for Experiment 4 using AVIRIS Data.

the classes are assumed to be exhaustive. As indicated in Table 4-1, the number of labeled samples is small. To retain enough test samples, only about 200 training samples are chosen for each class. Because of the limited labeled sample size, to obtain reasonably good initial estimates for comparing the EM and REM algorithms, the numbers of spectral channels are selected at 10, 20, 50, 67, and 100. These spectral features are again chosen by sampling the spectral channels at fixed intervals. Figure. 4-12 shows the classification results at the selected dimensions.

The results show that the REM algorithm performs better than the ML and EM methods. This demonstrates that although it is assumed that the scene contains no outliers, there are some outlying pixels that were not identified. This further justifies the motivation of using a robust parameter estimation method for the mixture model. The results also show that all methods exhibit the Hughes phenomenon. As discussed previously, the decline in performance at high

216

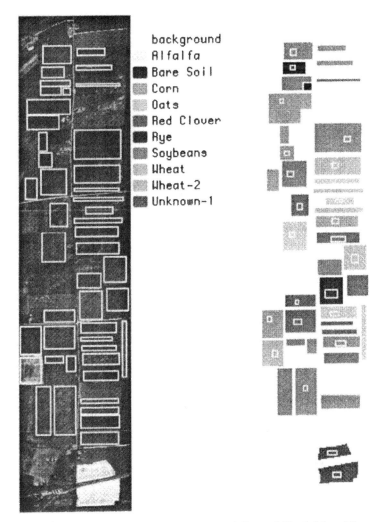

Figure 4-13. Flightline C1 data in image space and Ground Truth Map. The training (small rectangles) and test areas (large rectangles) used are shown. (In color on CD)

dimensionality is caused by the limited number of unlabeled samples available in the data set.

<u>Illustration 5</u>

This illustration is conducted using the data set designated Flightline C1 (FLC1) again. The data in image space and the ground truth map are shown in Figure 4-13. The training fields are marked in the ground truth map. The number of labeled samples and training samples in each class is shown in Table 4-2. The parameters are estimated using the training samples alone, the EM algorithm with various threshold settings, and the REM algorithm. For the EM algorithm, two chi-square threshold values (1% and 5%) are applied for comparison. The classification results are plotted in Figure 4-14.

Class Names	No. of Labeled Samples	No. of Training Samples
Alfalfa	3375	156
Bare Soil	1230	90
Corn	10625	331
Oats	5781	306
Red Clover	12147	614
Rye	2385	408
Soybeans	25133	631
Wheat	7827	340
Wheat-2	2091	120
Unknown-1	4034	322

Table 4-2. Class Description for Flightline C1 Data.

The entire Flightline C1 data set contains classes with a few pixels such as rural roads, farmsteads, and water that are not included in the training set. There may be other unknown classes that are not identified in the ground truth information. Therefore it is highly

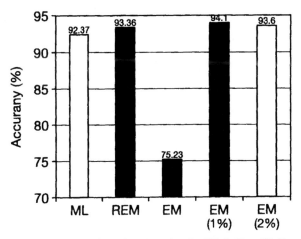

Figure 4-14. Classification Results for Flightline C1 Data.

likely that statistical outliers are present in the image. This is confirmed by experimental results. The performance of the EM algorithm is significantly lower than those of ML, REM, and EM with thresholding. Again, the experiment demonstrates that REM has similar performance as EM with thresholding, but without the need of setting a threshold.

4.7 Some Additional Comments

The problem of successfully analyzing a multispectral data set comes down to achieving an accurate statistical model of the data set in which the desired informational classes are separable components. Choosing training samples that accomplish this is not likely to produce a good model of the entire data set. This is because a multispectral image data set usually contains pixels of unknown classes that are either unknown or can be time-consuming to identify. The pixels of unknown origin may have density distributions quite different from the training classes and constitute statistical outliers. Without a list of exhaustive classes for the mixture model, the expectation maximization (EM) algorithm can converge to erroneous solutions because of the presence of statistical outliers. This problem necessitates a robust version of the EM algorithm that includes a measure of typicality for each sample. As a result a robust method for parameter estimation under the mixture model (REM) is useful for classifying multispectral data.

The experimental results have shown that the robust method performs better than the parameter estimation methods using the training samples alone (ML) and the EM algorithm in the presence of outliers. When no outliers are present, the EM and REM have similar performances and both are better than the ML approach. Specifically, when there are many unlabeled samples available, the EM and REM algorithms can mitigate the Hughes phenomenon, since they utilize unlabeled samples in addition to the training samples. When the number of unlabeled samples is limited, both EM and REM methods exhibit the Hughes phenomenon but still achieve better classification accuracy than the ML approach at lower dimensionality. Despite the promising results, the REM algorithm has several limitations. Since the weight function in the REM algorithm is based on class statistics, the initial parameter estimates are important in determining the convergence. In particular, a good covariance estimate requires a sufficient number of training samples. When the number of training samples is close to or less than the dimensionality, the covariance estimate becomes poor or singular and the EM or REM algorithm cannot be applied. This necessitates the use of a covariance estimation method for limited training samples. Such a scheme will be introduced next.

4.8 A Small Sample Covariance Estimation Scheme

As has been seen, it is frequently useful to use the class-conditional Gaussian density function

$$p(\mathbf{x} \mid \omega_i) = (2\pi)^{-N/2} |\Sigma_i|^{-1/2} \exp\left\{-\frac{1}{2}(\mathbf{x} - \mu_i)^T \Sigma_i^{-1}(\mathbf{x} - \mu_i)\right\}$$

as the discriminant function in multispectral analysis problems, and this necessitates the estimation of μ_i and Σ_i for each of the classes. Since it is frequently the case in remote sensing that there is a paucity of training samples by which to carry out this estimation, significant estimation error can result. This problem becomes more serious as the number of features increases, such as in the case of hyperspectral data.

Basically the problem is one of complexity and the number of parameters that must be estimated with a finite number of samples. The discriminant function above uses both the mean vector, μ_i and the covariance matrix, Σ_i. The advantage is that this allows for decision boundaries that are second order surfaces or with the use of several subclasses, a combination of segments of second-order surfaces. The minimum distance classifier, which utilizes only class mean values, μ_i, allows for decision boundaries that are only first order. To adequately define the location and shape of the more complex decision boundaries, more parameters, in the form of the elements of Σ_i, must be estimated. As the number of dimensions goes up, the number of parameters in the μ_i and especially in Σ_i increases very rapidly. The need to estimate more and more parameters with a fixed number of samples leads to increasing inaccuracy. As seen from the Hughes phenomenon, the detrimental effect induced by this inaccuracy eventually overcomes the advantage of the additional features.

As it turns out, with limited training samples, discriminant functions that do not contain so many parameters may provide improved classification results even though they result in simpler decision boundaries. We will next examine a covariance estimator that provides this advantage. This covariance estimator examines mixtures of the sample covariance, diagonal sample covariance, common covariance, and diagonal common covariance. Let us examine this idea in more detail.

$$\hat{\mu}_i = \frac{1}{N_i} \sum_{j=1}^{N_i} x_{ij}$$

The mean vector is typically estimated by the sample mean

$$\hat{\Sigma}_i = \frac{1}{N_i - 1} \sum_{j=1}^{N_i} (x_{ij} - \hat{\mu}_i)(x_{ij} - \hat{\mu}_i)^\mathsf{T}$$

where x_{ij} is sample j from class i. The covariance matrix is typically estimated by the (unbiased) sample covariance

or the maximum likelihood covariance estimate

$$\hat{\Sigma}_i^{ML} = \frac{1}{N_i} \sum_{j=1}^{N_i} (x_{ij} - \hat{\mu}_i)(x_{ij} - \hat{\mu}_i)^\mathsf{T}$$

The classification rule that results from using the class conditional maximum likelihood estimates for the mean and covariance in the discriminant function as if they were the true mean and covariance achieves optimal classification accuracy only asymptotically as the number of training samples increases toward infinity, and therefore the estimation error approaches zero. This classification scheme is

[40] T.W. Anderson, *An introduction to multivariate statistical analysis*, 2nd Ed., New York, Wiley, 1984, p. 209.

[41] J.H. Friedman, Regularized discriminant analysis, *J. of the American Statistical Association*, Vol. 84, pp. 165-175, March 1989.

not optimal when the training sample is finite.[40] When the training set is small, the sample estimate of the covariance is usually highly elliptical and can vary drastically from the true covariance. In fact, for p features, when the number of training samples is less than p+1, the sample covariance is always singular.

For limited training data the common covariance estimate, obtained by assuming all classes have the same covariance matrix, can lead to higher accuracy than the class conditional sample estimate, even when the true covariance matrices are quite different.[41] This leads to several possible assumptions that could turn out to be advantageous. To illustrate the possibilities, suppose that one has a 5 class problem and 6 dimensional data. Then the various possibilities for coefficients that must be estimated are illustrated by the following diagrams of covariance matrices:

Class 1	Class 2	Class 3	Class 4	Class 5
b	b	b	b	b
a b	a b	a b	a b	a b
a a b	a a b	a a b	a a b	a a b
a a a b	a a a b	a a a b	a a a b	a a a b
a a a a b	a a a a b	a a a a b	a a a a b	a a a a b
a a a a a b	a a a a a b	a a a a a b	a a a a a b	a a a a a b

Common Covar.
d
c d
c c d
c c c d
c c c c d
c c c c c d

Figure 4-15. Covariance coefficients that must be estimated for a 5 class, 6 dimensional problem

Now, for limited numbers of training samples, the greater the numbers of coefficients one must estimate, the lower is the accuracy of the estimates. If one were to attempt a normal estimate of individual class covariances, then one must estimate coefficients in positions marked a and b in Figure 4-15. If, on the other hand, it appears advantageous to ignore the correlation between channels, then one would only need to estimate coefficients marked b. If one were willing to assume that all classes have the same covariance, then one would only need to estimate the smaller number of coefficients marked c and d above. And finally, if in addition, one were willing to ignore the correlation between channels of this common covariance, one would only need to estimate the coefficients marked d above.

For p dimensional data, the number of coefficients in a class covariance function is $(p+1)p/2$. Table 4-3 illustrates the number of coefficients in the various covariance matrix forms that must be estimated for the case of 5 classes and several different numbers of features, p.

No. of Features p	Class Covariance (a & b above) $5\{(p+1)p/2\}$	Diagonal Class Covariance (b above) 5p	Common Covariance (c & d above) $\{p+1)p/2\}$	Diagonal Common Covariance (d above)
5	75	25	15	5
10	275	50	55	10
20	1050	100	210	20
50	6375	250	1275	50
200	100,500	1000	20,100	200

Table 4-3. The number of covariance matrix coefficients that must be estimated for various classifiers

If, for example, one has 100 training samples for each of 5 classes and 20 features, then, for the individual class covariance case, one might expect an accuracy problem attempting to estimate the 1050 coefficients with only 500 total training pixels. It is useful in any given case to determine whether the sample estimate or the common covariance estimate would be more appropriate in a given situation. This illustrates the manner in which a properly chosen estimator could improve classifier performance.

The estimator discussed here examines the sample covariance and the common covariance estimates, as well as their diagonal forms, to determine which would be most appropriate. Furthermore, it examines the following pairwise mixtures of estimators:

- sample covariance-diagonal sample covariance,
- sample covariance-common covariance, and
- common covariance-diagonal common covariance.

The estimator has the following form:

$$C_i(\alpha_i) = \begin{cases} (1-\alpha_i)\text{diag}(\Sigma_i) + \alpha_i\Sigma_i & 0 \le \alpha_i \le 1 \\ (2-\alpha_i)\Sigma_i + (\alpha_i - 1)S & 1 \le \alpha_i \le 2 \\ (3-\alpha_i)S + (\alpha_i - 2)\text{diag}(S) & 2 \le \alpha_i \le 3 \end{cases} \quad (4\text{-}7)$$

where $S = \dfrac{1}{L}\sum_{i=1}^{L}\Sigma_i$ is the common covariance matrix, computed by averaging the class covariances across all classes. This approach assumes that all classes have the same covariance matrix. The variable α_i is a mixing parameter that determines which mixture is selected.

To understand the decision boundary implication of the various values of α_i, recall that the discriminant function for the more general quadratic classifier is

$$g_i(\mathbf{X}) = \ln\frac{p^2(\omega_i)}{|\Sigma_i|} - (\mathbf{X} - \mu_i)^T \Sigma_i^{-1}(\mathbf{X} - \mu_i)$$

- If $\alpha_i = 0$, the diagonal sample covariance is used. Thus Σ_i is assumed to be diagonal, meaning that only the b's above are used. This is equivalent to assuming that the features are uncorrelated from feature to feature, that the classes are hyperelliptically distributed in feature space, and that a linear decision surface is produced.

- If $\alpha_i = 1$, the estimator returns the sample covariance estimate. Thus estimates of both the a's and the b's above are used and the decision surface would be a second-order surface. This is referred to as a quadratic classifier.

- If $\alpha_i = 2$, the common covariance is selected. In this case the discriminant function becomes

$$g_i(\mathbf{X}) = \ln\frac{p^2(\omega_i)}{|\mathbf{S}|} - (\mathbf{X} - \mu_i)^T \mathbf{S}^{-1}(\mathbf{X} - \mu_i)$$

This implies estimating the c's and d's above, that all classes have the same covariance matrix, and a linear decision surface would again result. However, now one has all of the samples of all classes with which to estimate one covariance matrix. This is referred to as the Fisher Linear Discriminant algorithm.

- If $\alpha_i = 3$, the diagonal common covariance results. In this case the discriminant function becomes

$$g_i(\mathbf{X}) = \ln\frac{p^2(\omega_i)}{|\mathrm{diag}(\mathbf{S})|} - (\mathbf{X} - \mu_i)^T [\mathrm{diag}(\mathbf{S})]^{-1}(\mathbf{X} - \mu_i)$$

Although the variances in all features are estimated, correlation among features is assumed to be zero, and a linear decision surface will result which is the perpendicular bisector of the line between the two class mean values. The location along this line between the means would be determined by the relative size of the class variances. If all the variances were the same such that the decision boundary was at the midpoint, this would be the same as the Minimum Distance classifier.

Other values of α_i lead to mixtures of two estimates.

The value of the mixing parameter is selected so that the best fit to the training samples is achieved, in the sense that the average likelihood of left-out samples is maximized. The technique is to remove one sample, estimate the mean and covariance from the remaining samples, then compute the likelihood of the sample that was left out, given the mean and covariance estimates. Each sample is removed in turn, and the average log likelihood is computed. Several mixtures are examined by changing the value of α_i, and the value that maximizes the average log likelihood is selected.

The mean of class i, without sample k, is

$$\hat{\mu}_{ilk} = \frac{1}{N_i - 1} \sum_{j=1 \atop j \neq k}^{N_i} x_{ij}$$

where the notation $i|k$ indicates the quantity is computed without using sample k from class i. The sample covariance of class i, without sample k, is

$$\hat{\Sigma}_{ilk} = \frac{1}{N_i - 2} \sum_{j=1, j \neq k}^{N_i} (\mathbf{x}_{ij} - \hat{\boldsymbol{\mu}}_{ilk})(\mathbf{x}_{ij} - \hat{\boldsymbol{\mu}}_{ilk})^T \qquad (4\text{-}8)$$

and the common covariance, without sample k, is

$$\hat{\mathbf{S}}_{ilk} = \left(\frac{1}{L} \sum_{i=1, i \neq j}^{L} \hat{\Sigma}_i \right) + \frac{1}{L} \hat{\Sigma}_{ilk}$$

The estimate for class i, without sample k, can then be computed as follows:

$$\mathbf{C}_{ilk}(\alpha_i) = \begin{cases} (1 - \alpha_i)\text{diag}(\hat{\Sigma}_{ilk}) + \alpha_i \hat{\Sigma}_{ilk} & 0 \leq \alpha_i \leq 1 \\ (2 - \alpha_i)\hat{\Sigma}_{ilk} + (\alpha_i - 1)\hat{\mathbf{S}}_{ilk} & 1 \leq \alpha_i \leq 2 \\ (3 - \alpha_i)\hat{\mathbf{S}}_{ilk} + (\alpha_i - 2)\text{diag}(\hat{\mathbf{S}}_{ilk}) & 2 \leq \alpha_i \leq 3 \end{cases}$$

Next the average log likelihood of $\mathbf{x}_{ij}$, is computed as follows:

$$B_i = \frac{1}{N_i} \sum_{k=1}^{N_i} \ln[p(\mathbf{x}) \mid \hat{\boldsymbol{\mu}}_{ilk}, \mathbf{C}_{ilk}(\alpha_i)]$$

This computation is repeated for several values of α_i over the range $0 \leq \alpha_i \leq 3$, and the value with the highest average log likelihood is selected. Once the appropriate value of α_i has been estimated, the proposed estimate is computed using all the training samples, and the result can be used in the maximum likelihood classifier.

Since evaluation of the Gaussian density function requires the inverse of the covariance matrix, an estimate of the covariance is only useful in classification if it is nonsingular (i.e., invertible). The sample covariance estimate is singular for p-dimensional data if there are fewer than p+1 samples available. Since a diagonal matrix is nonsingular if its diagonal elements are all nonzero, the proposed estimate is nonsingular as long as the sample covariance has nonzero diagonal elements, which is the usual case if there is more than one sample. Equation (4-8), however, requires the number of samples in each class to be at least three regardless of the dimension of the data. The proposed estimate, then, will usually be nonsingular with as few as three linearly independent samples per class.

This optimization criterion is referred to as the Leave One Out method, and this means for estimating $C(\alpha_i)$ is thus called the LOOC method. If implemented directly, the sample covariance and common covariance computations of this estimate would require computing the inverse and determinant of the (p x p) matrix $C_{i|k}$ for each training pixel, which would be quite expensive computationally. Fortunately a significant reduction in the required computation can be achieved that allows the determinant and inverse to be computed efficiently.[42] Further, an approximation is possible in carrying out the sample covariance-diagonal sample covariance and common covariance-diagonal common covariance computations that shorten the computation even further.[43] Taken together, these short cuts make the length of the computation involved quite practical.

4.9 Results for Some Examples

To illustrate the performance of this method, the results of several experiments with the estimator will be presented. First will be results

---

[42] J. P. Hoffbeck, and D. A. Landgrebe, Covariance estimation and classification with limited training data, *IEEE Transactions on Pattern Analysis and Machine Intelligence*, Vol. 18, no. 7, pp. 763-767, July 1996.

[43] Ibid.

from computer-generated data in order to show how the estimator performs in known situations. This is followed by results in the case of two real data sets. In all cases the training set size is limited and the classification accuracy resulting from LOOC will be compared with several different covariance estimates (in decreasing order of complexity of the decision boundaries resulting): sample covariance, common covariance, Euclidean distance.

For the compute-generated data set experiments, 15 independent random training samples were drawn from three different normal distributions, the appropriate statistics were estimated, and the classification accuracy was measured by classifying 100 independent test samples per class. Four experiments with various distributions adapted from Friedman[44] were performed using four different dimensionalities. Each experiment was repeated 25 times, and the mean and standard deviation of the classification accuracy were recorded. In all the experiments, α_i took the values (0.0, 0.25, 0.50, 0.75, 1.0, 1.25, 1.50, 1.75, 2.0, 2.25, 2.50, 2.75, 3.00).

Figure 4-16 shows the results for the case where all three classes have an identity covariance matrix but slightly different mean values. The mean of the first class was at the origin, the mean of the second class was 3.0 in the first variable and zero in the other variables, and the mean of the third class was 3.0 in the second variable and zero in the other variables. In this case one would expect the Euclidean distance classifier to perform best, and this is the case. The LOOC method provided results that were nearly as

---

[44] J.H. Friedman, Regularized discriminant analysis, *Journal of the American Statistical Association*, Vol. 84, pp. 165-175, March 1989. This paper presents another means for estimating statistics in such circumstances. It provides similar results to the current method, but takes substantially more computation.

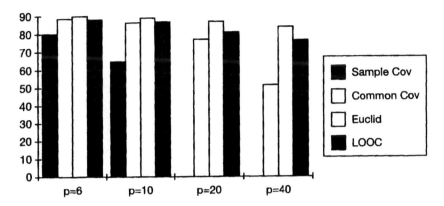

Figure 4-16. Results for the first experiment, where all classes had an identity covariance matrix.

good, however. Note that, since there were only 15 training samples, the Sample Covariance method could not be used for p > 14.

The results for the second experiment are shown in Figure 4-17. In this case, the classes have different means and different covariances.

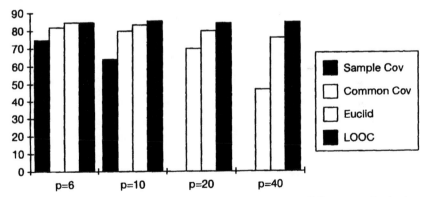

Figure 4-17. Results where the three classes have different, spherical covariance matrices and different mean vectors. The covariance of class one, two, and three was I, 2I, and 3I, respectively. The mean of the first class was at the origin, the mean of the second class was 3.0 in the first variable and zero in the other elements, and the mean of the third class was 4.0 in the second variable and zero in the other elements.

Thus, neither the common covariance nor the Euclidean distance classifier would be optimal, and their performance would be expected to fall off as the dimensionality increased. The LOOC approach supplied consistent results, even when the dimensionality substantially exceeded the number of training samples.

The results for the third experiment are given in Figure 4-18. Here the classes have different means and covariances again, but this time the three classes have highly elliptical distributions. In this case the performance of all four methods decreases with increasing dimensionality, but the decline for LOOC is more moderate.

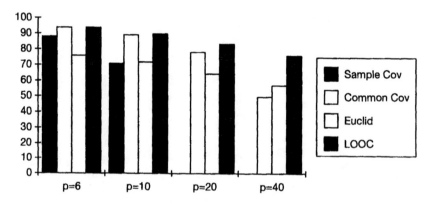

Figure 4-18. Results for the third experiment where the three classes have the same highly elliptical covariance matrix, and the primary difference in the mean vectors is in the variables with low variance. The covariance matrix for all three classes is a diagonal matrix whose diagonal elements

are given by $\sigma_i = \left[\dfrac{9(i-1)}{(p-1)} + 1 \right]^2$ $1 \le i \le p$. The mean vector of the first class

is at the origin; the elements of the mean vector of the second class are

given by $\mu_{2,i} = 2.5 \sqrt{\dfrac{\sigma_i}{p} \left(\dfrac{p-i}{p/2 - i} \right)}$ $1 \le i \le p$, and the mean of class

three is defined by $\mu_{3,i} = (-1)^i \mu_{2,i}$.

The results for the fourth experiment are given in Figure 4-19. Here the classes have the same highly elliptical distributions as in experiment 3, but this time the relationship between the means and their covariances has been changed. In this case the performance of all four methods again decrease with increasing dimensionality, but the decline for the Euclidean distance method is more moderate. The two following experiments were performed on data taken in 1992 by the Airborne Visible/Infrared Imaging Spectrometer (AVIRIS). This instrument captures data of the Earth's surface in 220 spectral bands covering the range 0.4 to 2.5 μm. In order to conduct the experiments, several samples (pixels) of various ground cover classes were identified in each of the scenes. Then a small percentage of the samples were selected at random and used to estimate the appropriate statistics. Finally, the remaining pixels were

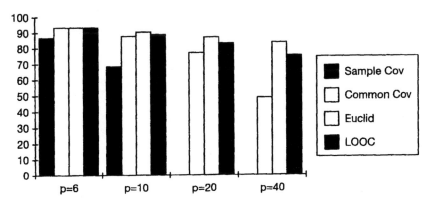

Figure 4-19. Results for the fourth experiment. The same highly elliptical covariance matrix from experiment 3 is again used for all three classes, but the difference in mean vectors occurs in the variables that have high variance. The mean of the first class is again at the origin, the mean of class two is defined by

$$\mu_{2,i} = 2.5 \sqrt{\frac{\sigma_i}{p}} \left(\frac{i-1}{p/2 - i} \right) \quad 1 \le i \le p, \text{ and the mean of class}$$

three is defined as $\mu_{3,i} = (-1)^i \mu_{2,i}$.

233

classified to measure the classification accuracy. The experiment was repeated 10 times, and the mean and standard deviation were recorded. The experiments were conducted with four different numbers of features. The features were selected evenly spaced across the spectrum but did not include those bands that lie in the water absorption regions.

For experiment 5, data from a Cuprite Nevada site was used. This site has only sparse vegetation and several exposed minerals. Four of the classes (alunite, buddingtonite, kaolinite, and quartz) had easily identifiable absorption features and were labeled by

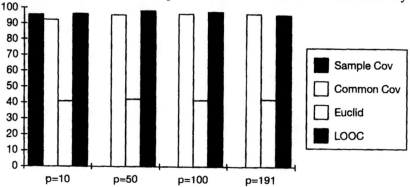

Figure 4-20. Results using AVIRIS data from the Cuprite, Neada, site.

45 A.F.H. Goetz, and V. Srivastava, "Mineralogical Mapping in the Cuprite Mining District, Nevada," *Proceedings of the Airborne Imaging Spectrometer Data Analysis Workshop*, JPL Publication 85-41, pp. 22-31, April 1985.

46 M.J. Abrams, R.P. Ashley, L.C. Rowan, A.F.H. Goetz, and A.B. Kahle, "Mapping of Hydrothermal Alteration in the Cuprite Mining District, Nevada, Using Aircraft Scanner Images for the Spectral Region 0.46 to 2.36 μm," *Geology*, Vol. 5, pp. 713-718, December 1977.

47 F.A. Kruse, K.S. Kierein-Young, and J.W. Boardman, "Mineral Mapping at Cuprite, Nevada with a 63-Channel Imaging Spectrometer," *Photogrammetric Engineering and Remote Sensing*, Vol. 56, pp. 83-92, January 1990.

comparing the log residue spectrum to laboratory reflectance curves.[45] The classes alluvium, argillized, tuff, and playa were identified by comparing the scene to a geology map produced from ground observations.[46, 47] A total of 2744 samples (pixels) and 191 bands (0.40-1.34, 1.43-1.80, 1.96-2.46μm) were available for the experiment. The number of training samples in each class was 145, 14, 46, 77, 137, 50, 58, and 18, which represented 20% of the total number of available labeled samples. The results of experiment 5 are presented in Figure 4-20. In this case it is difficult to describe

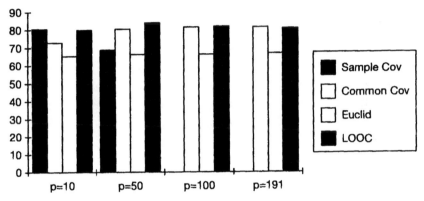

Figure 4-21. Results for AVIRIS data over an agricultural site.

the relationships among the mean vectors and covariance matrices as is typical with real data. However, both the Common Covariance classifier and the LOOC method provided good results at all dimensionalities. Since three of the classes had fewer than 51 samples, the Sample Covariance method could be used only in the case of p = 10 here.

For the sixth experiment, data from an agricultural site in Northwestern Indiana referred to as Indian Pine was used. In this case ground observations were used to identify a total of 2521 samples from the following classes: soybeans with no residue, soybeans with corn residue, corn with no residue, corn with soybean residue, corn with wheat residue, and wheat with no residue. From the total number of available labeled samples, 20% were used as

235

training samples, making the number of training samples in each class 104, 90, 74, 98, 77, and 60. The results for the sixth experiment are shown in Figure 4-21. Again the Common Covariance and the LOOC methods held up across all dimensionalities.

The purpose of these experiments is to display the performance of the classifiers in a variety of circumstances when the training sets were limited. The number of training samples used was intended to place the experiment just above the threshold, under which satisfactory results cannot be obtained by any method but below the number for which selection of the scheme can be to routinely use the Sample Covariance method. This approach tends to be the standard approach for conventional multispectral ($p \leq 10$ or so) data. In this circumstance, it is clear that the most complex decision boundary does not necessarily give the best results, but from a practical standpoint the user does not have any convenient way to determine what will work best. An advantage of LOOC, then, is that this algorithm makes a good decision for the user.

The potential for hyperspectral data is for it to make possible discrimination between much more detailed classes than would be possible with conventional multispectral data, but the typical shortage of training samples becomes an increasingly serious matter as the dimension is increased to achieve this potential. The LOOC is a means to mitigate this problem. It was seen that it consistently provided good results in the wide variety of examples presented here.

5 Hyperspectral Data Characteristics

This chapter is devoted to matters that arise when the number of spectral bands of a data set become large, i.e., 10's to 100's. It begins with a means for visualization of high dimensional statistics, and then proceeds to discuss some of the nonintuitive aspects of high dimensional feature spaces. There is a discussion of the role of second order statistics relative to first order statistics in discrimination between classes. All of this is a prelude to developing sound means for analyzing hyperspectral data sets, which are presented in the following chapters.

5.1 Introduction

It is common to speak of remote sensing data as "imagery" and indeed, the most common way for human view of such data is in image form. As one begins to think of ordinary color imagery as multispectral data, one begins to think of the three spectral bands that are used to make up the color image, as shown in Figure 5-1. However, multispectral data usually has more than just three bands.

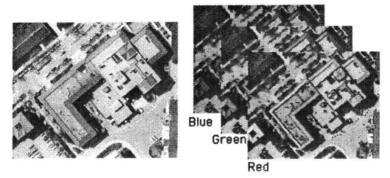

Figure 5-1. A color image and the three separations of it showing the blue, green, and red separations that make it up. Notice, for example, that the small square building with a red rooftop is brighter in the red than the blue and green separations. (In color on CD)

Advancements in solid-state technology in recent years have made possible collection of multispectral data in as many as 200 or more spectral bands. Such data are referred to as hyperspectral data. Indeed, it is no longer reasonable to speak of it as "imagery," as to do so is misleading. One can "see" only a small part of the information-bearing variations of hyperspectral data in image form, the arrangement of the data in three of the many bands that it is composed of. Thus, though using three of the large number of bands it contains is still useful in locating a pixel geographically, one needs new tools in being able to perceive and manipulate the data for analytical purposes.

The motivation for higher dimensionality is that it should make possible discrimination among a much larger set of classes, and classes that are more detailed in nature. However, the increase in data complexity in this case makes it even more important that proper signal models be used and that care be exercised in perceptions and assumptions about what occurs in these higher dimensional feature spaces.

On the intuition in higher dimensional space, consider the following:[48]

- Borsuk's Conjecture: In higher-dimensional space, if you break a stick in two, both pieces are shorter than the original.

- Keller's Conjecture: It is possible to use cubes (hypercubes) of equal size to fill an n-dimensional space, leaving neither overlaps nor underlaps.

---

[48]   Science, Vol. 259, 1 Jan. 1993, pp. 26-27

Counterexamples to both have been found for higher dimensional spaces. The lesson to be drawn from this is that geometric concepts that are self-evident in two- or three-dimensional space do not necessarily apply in higher dimensional space. We will explore this point in greater detail shortly.

5.2 A Visualization Tool

Before doing so, however, we note that one approach to dealing with high-dimensional data would be to develop visualization tools to show what is taking place in such spaces, tools that provide information in both qualitative and quantitative form in order for the analyst to make good choices about how to proceed. As an example of the problem, shown in Table 5-1 is the correlation matrix for a sample of 12-dimensional data. With practice, one may be able to become adept at viewing such a matrix and gain perceptions about the sample involved. For example, it is seen that the bands in the visible region (channels 1–10) tend to be positively correlated with one another, but that visible bands tend to be negatively correlated with those in the reflective infrared region, a relationship common of vegetation. However, if one increases the dimensionality by an order of magnitude or even less, drawing any kind of perceptive knowledge in such a fashion would become impossible.

Band µm		1	2	3	4	5	6	7	8	9	10	11	12
1	0.40-0.44	1.0											
2	0.44-0.46	0.66	1.0										
3	0.46-0.48	0.75	0.69	1.0									
4	0.48-0.50	0.64	0.75	0.69	1.0								
5	0.50-0.52	0.68	0.70	0.79	0.71	1.0							
6	0.52-0.55	0.31	0.45	0.45	0.52	0.63	1.0						
7	0.55-0.58	0.41	0.53	0.57	0.61	0.60	0.71	1.0					
8	0.58-0.62	0.61	0.75	0.71	0.80	0.82	0.67	0.67	1.0				
9	0.62-0.66	0.58	0.75	0.67	0.81	0.73	0.59	0.66	0.89	1.0			
10	0.66-0.72	0.34	0.50	0.52	0.57	0.64	0.79	0.74	0.72	0.67	1.0		
11	0.72-0.80	-0.45	-0.50	-0.38	-0.51	-0.32	0.07	-0.11	-0.47	-0.59	-0.02	1.0	
12	0.80-1.0	-0.58	-0.58	-0.56	-0.57	-0.48	-0.01	-0.20	-0.52	-0.59	-0.10	0.75	1.0

Table 5-1. A correlation matrix for a sample of 12-dimensional data.

A visualization tool for providing this kind of perception in hyperspectral cases can be devised using color instead of numerical values using the following scheme.[49] Figure 5-2 shows a legend for the relationship between color and correlation value.

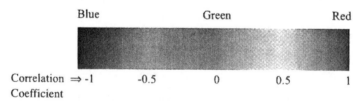

Figure 5-2. Color-to-correlation value legend. (In color on CD)

Figure 5-3 shows a presentation of perceptional information about two classes in a 200-dimensional space. These will be referred to as Statistics Images. In the upper portion is presented information similar to that in Table 5-1, but in terms of color rather than numerical values. From these it is easy to see that the two classes have different second order statistics, implying that they may well

[49] Chulhee Lee and D. A. Landgrebe, Analyzing high dimensional multispectral data, *IEEE Transactions on Geoscience and Remote Sensing,* Vol. 31, No. 4, pp. 792-800, July 1993.

be separable in that space. The lower portion of these presentations shows a more conventional graph of the data in spectral space, showing the mean value ± σ.

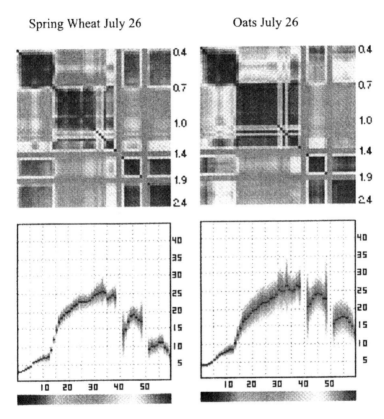

Figure 5-3. Two spectral classes using color displays of class statistics to aid is providing comparative class information. (In color on CD)

5.3 Accuracy vs. Statistics Order

To explore further, in thinking of models for a signal in N-dimensional feature space, we have pointed out that the first-order statistic, i.e. the class mean, provides the information about *where* in the space *the class is located*, while the second-order statistics, i.e., the variances and covariances, describe *its shape*. There is evidence to suggest in a general sense that for simple signals, the location information provided by the first order statistics is often more important in the discrimination process. However, as the dimensionality and signal complexity grows, the shape information provided by the second order statistics can in some cases begin to dominate, and the location of the distributions becomes relatively

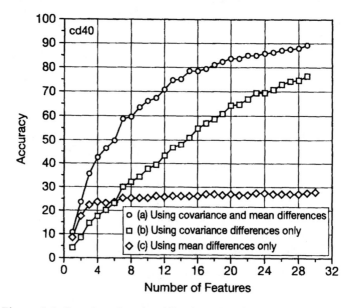

Figure 5-4. Results of a classification showing accuracy vs. number of features used. Shown are the accuracies for use of both first- and second-order statistics, second-order statistics only and first-order statistics only. Notice the dominance of second-order statistics for higher dimensions.

less important. This is illustrated by the research result shown in Figure 5-4.

For this experiment a multispectral data set with a large number of spectral bands was analyzed using standard pattern recognition techniques. The data were classified first by a single spectral feature, then by two, and continuing on with greater and greater numbers of features. Three different classification schemes were use:, (a) a standard maximum likelihood Gaussian scheme which included both the means and the covariance matrices, i.e., both first and second order variations, (b) the same except with all classes adjusted to the same mean value, so that the classes differed only in their covariances, and (c) a minimum distance to means scheme such that mean differences are used, but covariances are ignored.

It is seen from the results in Figure 5-4 that case (a) produced clearly the best result, as might be expected. In comparing cases (b) and (c) it is seen that at first the classifier using mean differences performed best. However, as the number of features was increased, this performance soon became saturated, and improved no further. On the other hand, while the classifier of case (b) which used only second-order effects, was at first the poorest, it soon outperformed that of case (c) and its performance continued to improve as ever greater numbers of features were used. Thus it is seen that second-order effects, in this case represented by the class covariances, are not particularly significant at low dimensionality, but they become so as the number of features grows, to the point where they become much more significant than the mean differences between classes at any dimensionality.

243

Various indications of the significance of second order variations have occurred over time. Consider the result shown in Figure 5-5.[50] In an attempt to learn about the sensitivity of multispectral data analysis to spatially misregistered spectral bands, Thematic Mapper (7 band, 30 m spatial resolution) data were simulated with (12 band, 3 m) aircraft data by averaging pixels and bands. In the process, varying amounts of spatial misregistration of the bands were included in the simulated data sets. The data sets were then classified and the number of pixels whose classification changed from the baseline (nonmisregistered) data set noted. Figure 5-5 shows one such result.

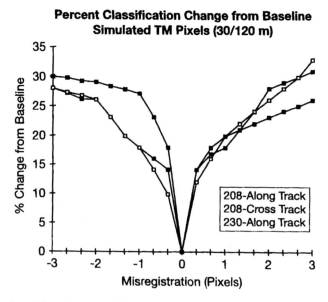

Figure 5-5. Classification change vs. misregistration for simulated TM data.

50 P. H.Swain, V. C.Vanderbilt, and C. D. Jobusch, A quantitative applications-oriented evaluation of hematic Mapper design specifications, *IEEE Transactions on Geoscience and Remote Sensing,* Vol. GE-20, No. 3, pp. 370-377, July 1982.

Figure 1-4. A color air view of the scene of Figure 1-3.

Figure 1-5. The scene of Figures 1-3 and 1-4 but in color IR form.

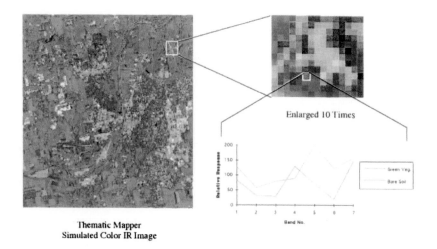

Enlarged 10 Times

Thematic Mapper
Simulated Color IR Image

Figure 1-8. The multispectral concept, showing the signal from a pixel expressed as a graph of response vs. spectral band number.

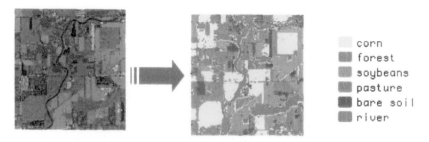

corn
forest
soybeans
pasture
bare soil
river

Figure 1-9. A thematic map of an agricultural area created from Thematic Mapper multispectral data.

Figure 3-1a. A "Picture." Pixels are a few centimeters in size. Objects are readily identifiable by human view. Viewing the area of Figure 3-1b at this resolution would result in about 10^6 times as much data.

Figure 3-1b. A Landsat image. Pixels are 30 meters in size. Objects are less identifiable by human view, but the volume of data is much more manageable.

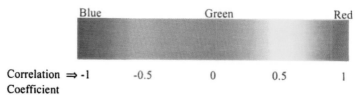

Blue   Green   Red

Correlation ⇒ -1   -0.5   0   0.5   1
Coefficient

Figure 5-2. Color-to-correlation value legend.

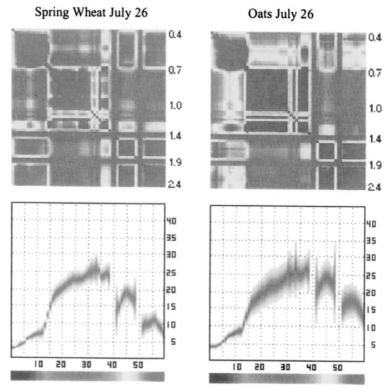

Spring Wheat July 26   Oats July 26

Figure 5-3. Two spectral classes using color displays of class
statistics to aid is providing comparative class information.

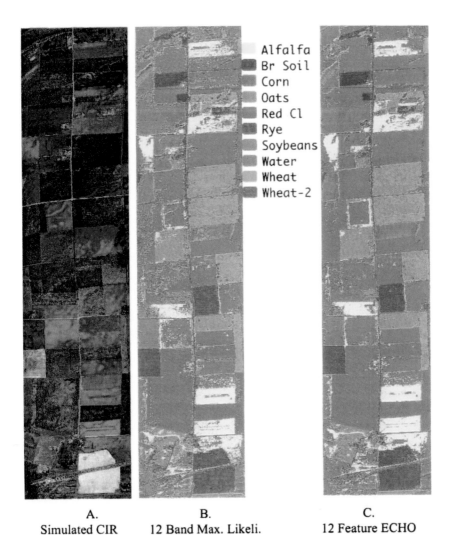

Alfalfa
Br Soil
Corn
Oats
Red Cl
Rye
Soybeans
Water
Wheat
Wheat-2

A.	B.	C.
Simulated CIR Image	12 Band Max. Likeli. Classification	12 Feature ECHO Classification using Enhanced Statistics

Figure 7-4. The data set in image form and results from analysis illustrating the effects of enhanced statistics.

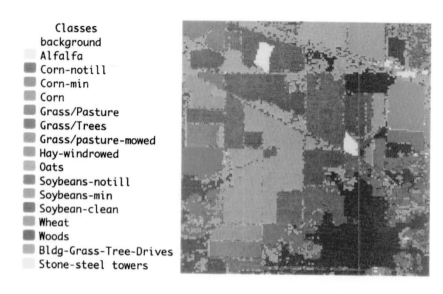

Classes
background
Alfalfa
Corn-notill
Corn-min
Corn
Grass/Pasture
Grass/Trees
Grass/pasture-mowed
Hay-windrowed
Oats
Soybeans-notill
Soybeans-min
Soybean-clean
Wheat
Woods
Bldg-Grass-Tree-Drives
Stone-steel towers

Figure 7-9 Thematic map resulting from the classification using
LOOC statistics and all 220 bands.

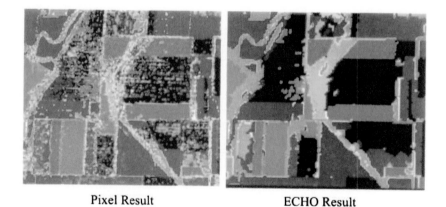

Pixel Result   ECHO Result

Figure 8-18. Thematic classification maps.

Groups
background
Roofs
Road
Grass
Trees
Trail
Water
Shadow

Figure 7-22. A simulated color IR image of a data set collected over the Washington DC Mall. Bands 60, 27, and 17 of 210 bands were used for this image space presentation.

Figure 7-23. Thematic map generated from the hyperspectral data set.

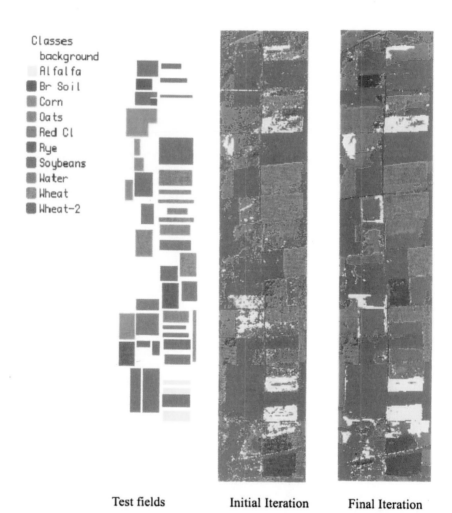

Classes
  background
  Alfalfa
  Br Soil
  Corn
  Oats
  Red Cl
  Rye
  Soybeans
  Water
  Wheat
  Wheat-2

Test fields   Initial Iteration   Final Iteration

Figure 7-26. Test fields and thematic map results for
FLC1.

The data used were from an agricultural area where the fields were much larger than the pixel size. Thus misregistration of as little as 0.3 pixel would not have moved many pixels from one field to another, i.e., moved a pixel from one class to another. Such a small change in registration would first show in the second order statistics, changing how the spectral response varies about its mean. This appears to be the explanation for the exceptional sensitivity to misregistration that shows up in this experiment.

5.4 High-Dimensional Spaces: A Closer Look

The increased dimensionality of such hyperspectral data presumably greatly enhances the data information content but provides a challenge to conventional techniques for analyzing such data. The move from what might be called conventional multispectral data, data having perhaps 10 bands or less, to such hyperspectral data with as many as several hundred bands is indeed a large jump. Quite aside from a potentially larger computational task, it is appropriate to ask if the increase in dimensionality does not result in fundamental changes that must be accommodated in the data analysis procedures to be used if all of the available information in the data is to be realized. Is it necessary to even revise how one thinks of the data and the analysis task?

We will begin such a study by reviewing some of the basic properties of high-dimensional spaces.[51] It will turn out that human experience in three-dimensional space tends to mislead one's intuition of

---

[51] Much of the material in this section is taken from L. Jimenez and D. Landgrebe, Supervised classification in high dimensional space: geometrical, statistical and asymptotical properties of multivariate data, *IEEE Transactions on Systems, Man, and Cybernetics,* Vol. 28 Part C, Number 1, pp. 39-54, February 1998.

geometrical and statistical properties in high-dimensional space, properties that must guide choices in the data analysis process. Using Euclidean and Cartesian geometry, in the following section high-dimensional space properties are investigated and their implications for high-dimensional data and their analysis is studied in order to illuminate the differences between conventional spaces and hyper-dimensional space.

The complexity of dimensionality has been known for more than three decades, and the impact of dimensionality varies from one field to another. In combinatorial optimization over many dimensions, it is seen as an exponential growth of the computational effort with the number of dimensions. In statistics it manifests itself as a problem with parameter or density estimation due to the paucity of data. The negative effect of this paucity results from certain geometrical, statistical, and asymptotical properties of high dimensional feature space. These characteristics exhibit surprising behavior of data in higher dimensions.

The conceptual barrier between our experience in three dimensional Euclidean space and higher dimensional spaces makes it difficult for us to have proper perception of the properties of high dimensional space and the consequences in high dimensional data behavior. There are assumptions about high dimensional space that we tend to relate to our three-dimensional space intuition, assumptions as to where the concentration of volume is in hyper-cubes, hyperspheres, and hyperellipsoids or where the data concentration is in density function families, such as normal and uniform, that are important for statistical purposes. Other important perceptions that are relevant for statistical analysis are, for example, how the diagonals relate to the coordinates, what numbers of labeled samples are required for supervised classification, when normality can be assumed in data, and how important is the mean and covariance difference in the process of discrimination among different statistical classes. Next some characteristics of high-dimensional space will be studied, and their impact in supervised classification data analysis will be discussed. Most of these properties do not fit our experience in three-dimensional Euclidean space as mentioned before.

Geometrical, Statistical, and Asymptotical Properties

In this section some unusual or unexpected hyperspace characteristics are illustrated. These illustrations are intended to show that higher-dimensional space is quite different from the three-dimensional space with which we are familiar.

A. The volume of a hypercube concentrates in the corners.[52]

As dimensionality increases it has been shown[53] that the volume of a hypersphere of radius r and dimension d is given by the equation:

$$V_s(r) = \frac{2r^d}{d} \frac{\pi^{\frac{d}{2}}}{\Gamma\left(\frac{d}{2}\right)}$$

The volume of a hypercube in $[-r, r]^d$ is given by the equation

$$V_c(r) = (2r)^d$$

The fraction of the volume of a hypersphere inscribed in a hypercube of the same major dimension is

$$f_{d1} = \frac{V_s(r)}{V_c(r)} = \frac{\pi^{d/2}}{d 2^{d-1} \Gamma\left(d/2\right)}$$

---

52 D. W. Scott, Multivariate Density Estimation, Wiley, 1992.
53 M. G. Kendall, A Course in the Geometry of n-Dimensions, Hafner Publishing, 1961.

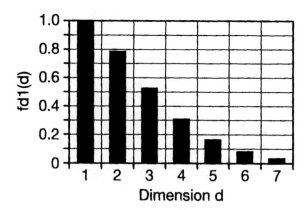

Figure 5-6. Fractional volume of a hypersphere inscribed in a hypercube as a function of dimensionality.

where d is the number of dimensions. Figure 5-6 shows how the ratio f_{d1} decreases as the dimensionality increases. Note that $\lim_{d\to\infty} f_{d1} = 0$, which implies that the volume of the hypercube is increasingly concentrated in the corners as d increases.

The Volume of a Hypersphere Concentrates in an Outside Shell [54, 55]

The fraction of the volume in a shell defined by a sphere of radius $r - \varepsilon$ inscribed inside a sphere of radius r is:

$$f_{d2} = \frac{V_d(r) - V_d(r - \varepsilon)}{V_d(r)} = \frac{r^d - (r - \varepsilon)^d}{r^d} = 1 - \left(1 - \frac{\varepsilon}{r}\right)^d$$

Figure 5-7 illustrates the case $\varepsilon = r/5$, where as the dimension increases, the volume concentrates in the outside shell. Note that ,

[54] D. W. Scott, Multivariate density estimation, Wiley, 1992.
[55] E. J. Wegman, Hyperdimensional data analysis using parallel coordinates, Journal of the American Statistical Association, Vol. 85, No. 411, pp. 664-675, 1990.

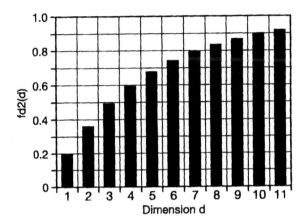

Figure 5-7. Volume of a hypersphere contained in the outside shell as a function of dimensionality for $\varepsilon = r/5$.

$$\lim_{d \to \infty} f_{d2} = 1 \text{ for all } \varepsilon > 0$$

implying that most of the volume of a hypersphere is concentrated in an outside shell.

The Volume of a Hyperellipsoid Concentrates in an Outside Shell.

Here the previous result will be generalized to a hyperellipsoid. Let the equation of a hyperellipsoid in d dimensions be written as

$$\frac{X_1^2}{\lambda_1^2} + \frac{X_2^2}{\lambda_2^2} + \cdots + \frac{X_d^2}{\lambda_d^2} = 1$$

The volume is calculated by the equation[56]

$$V_e(\lambda_i) = \frac{2 \prod_{i=1}^{d} \lambda_i}{d} \frac{\pi^{\frac{d}{2}}}{\Gamma\left(\frac{d}{2}\right)}$$

---

56 M. G. Kendall, A Course in the Geometry of n-Dimensions, Hafner Publishing, 1961.

The volume of a hyperellipsoid defined by the equation:

$$\frac{X_1^2}{(\lambda_1 - \delta_1)^2} + \frac{X_2^2}{(\lambda_2 - \delta_2)^2} + \cdots + \frac{X_d^2}{(\lambda_d - \delta_d)^2} = 1$$

where, $0 < \delta_i < \lambda_i$ for all i is calculated by:

$$V_e(\lambda_i - \delta_i) = \frac{2\prod\limits_{i=1}^{d}(\lambda_i - \delta_i)}{d} \frac{\pi^{\frac{d}{2}}}{\Gamma\left(\frac{d}{2}\right)}$$

The fraction of the volume of $V_e(\lambda_i - \delta_i)$ inscribed in the volume $V_e(\lambda_i)$ is:

$$f_{d3} = \frac{\prod\limits_{i=1}^{d}(\lambda_i - \delta_i)}{\prod\limits_{i=1}^{d}\lambda_i} = \prod\limits_{i=1}^{d}\left(1 - \frac{\delta_i}{\lambda_i}\right)$$

Let $\gamma_{min} = min\left(\frac{\delta_i}{\lambda_i}\right)$, then

$$f_{d3} = \prod\limits_{i=1}^{d}\left(1 - \frac{\delta_i}{\lambda_i}\right) \leq \prod\limits_{i=1}^{d}\left(1 - \gamma_{min}\right) = \left(1 - \gamma_{min}\right)^d$$

Using the fact that $f_{d3} \geq 0$, it is concluded that $\lim\limits_{d\to\infty} f_{d3} = 0$.

The characteristics previously mentioned have at least two important consequences for high dimensional data that appear immediately. The first is that

- High dimensional space is mostly empty. This implies that multivariate data in R^d is usually in a lower dimensional structure. As a consequence high dimensional data can be projected to a lower dimensional subspace without losing significant information in terms of separability among the different statistical classes.

The second consequence of the foregoing, is that

- Normally distributed data have a tendency to concentrate in the tails; similarly uniformly distributed data will likely be collected in the corners, making density estimation more difficult. Local neighborhoods are almost surely empty, producing the effect of losing detailed density estimation.

It is well known that the Gaussian density function is bellshaped and symmetrical about its mean. How can it be true that high-dimensional Gaussian data concentrates in the tails of the class, since it is clear from the Gaussian density function that the "most likely" values are near the mean, and not in the tails? This paradox can be explained as follows.[57] First note what happens to the magnitude of a zero mean Gaussian density function as the dimensionality increases. This is shown in Figure 5-8.

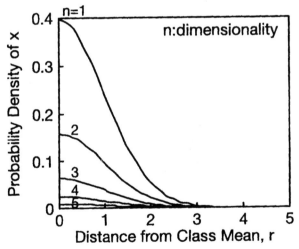

Figure 5-8. The magnitude of a Gaussian density vs. distance from the mean for various dimensionalities.

---

[57] This explanation was provided by graduate student Pi-fuei Hsieh.

It is seen that, while the shape of the curve remains bell-shaped, its magnitude becomes smaller with increasing dimensionality, as it must, since the overall volume must remain one, and, of course, it decreases exponentially as the radius r from the mean increases.

Next, consider how the volume density in the space changes as dimensionality increases. The volume of a hypersphere of radius r as a function of dimensionality is given by

$$V_s(r) = \text{volume of a hypersphere} = \frac{2r^d \; \pi^{d/2}}{d \; \Gamma(d/2)}$$

Therefore, the volume in a differential shell as a function of radius r is

$$\frac{dV}{dr} = \frac{2\pi^{d/2}}{\Gamma(d/2)} r^{(d-1)}$$

A plot for several values of d is given in Figure 5-9.

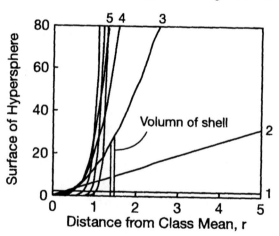

Figure 5-9. Differential surface volume of a hypersphere vs. distance from the center or mean for several dimensions.

Clearly, the volume available in a differential shell at radius r increases very rapidly with r as d becomes larger. Then the probability mass as a function of radius r, the combination of these two, may be shown to be

$$f_r(r) = \frac{r^{d-1}e^{-(r^2/2)}}{2^{(d/2)-1}\Gamma(d/2)}$$

This function is plotted in Figure 5-10. It may be shown that the peak of this function occurs at $\sqrt{d-1}$.

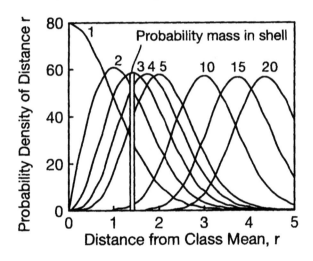

Figure 5-10. Probability mass vs. distance from the mean for a Gaussian density of several different dimensionalities.

Because the volume of a differential shell increases much more rapidly with r than the density function decreases, the net effect is as shown in the graph of Figure 5-10. Thus, it is seen that the peak of the probability mass moves away from the mean as the dimensionality increases, indicating that "most of the data becomes concentrated in the tails of the density" even though the data are Gaussianly distributed.

To further elucidate such specific multivariate data behavior, the following experiment was developed. Multivariate normally and uniformly distributed data were generated. The normal and uniform variables are independent identically distributed samples from the distributions $N(0, 1)$ and $U(-1, 1)$, respectively. Figures 5-11 and 5-12 illustrate the histograms of random variables, the distance from the zero coordinate and its square, which are functions of normal or uniform vectors for different number of dimensions.

These experiments show how the means and the standard deviations are functions of the number of dimensions. As the dimensionality increases the data concentrates in an outside shell. The mean and standard deviation of two random variables

$$r = \sqrt{\sum_{i=1}^{d} x_i^2} \quad \text{and} \quad R = \sum_{i=1}^{d} x_i^2$$

are computed. These variables are the distance and the square of the Euclidean distance of the random vectors. The values of the parameters and the histograms of the random variables shown in Figure 5-11 and 5-12 are for normal and uniform distributions of data. As the dimensionality increases, the distance from the zero coordinate of both random variables increases as well. Notice that the data have a tendency to concentrate in an outside shell and how the shell's distance from the zero coordinate increases with the increment of the number of dimensions.

Here $R = \sum_{i=1}^{d} x_i^2$ has a chi-square distribution with d degrees of freedom where the x_i's are samples from the $N(0,1)$ distribution. The mean and variance of R are $E(R) = d$, $Var(R) = 2d$.[58] This conclusion supports the previous thesis.

---

[58]   L. L.Scharf, Statistical Signal Processing. Detection, Estimation, and Time Series Analysis, Addison-Wesley, 1991, pp. 62-64.

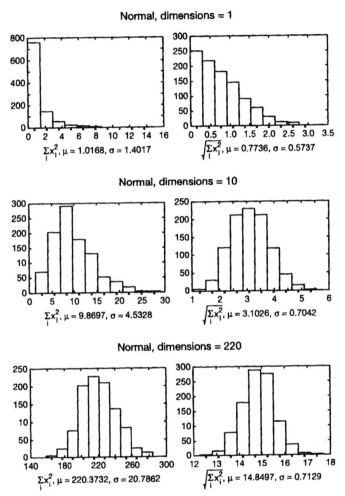

Figure 5-11. Histograms of functions of Normally distributed

Under these circumstances it would be difficult to implement any density estimation procedure and obtain accurate results. Generally nonparametric approaches will have even greater problems with high dimensional data.

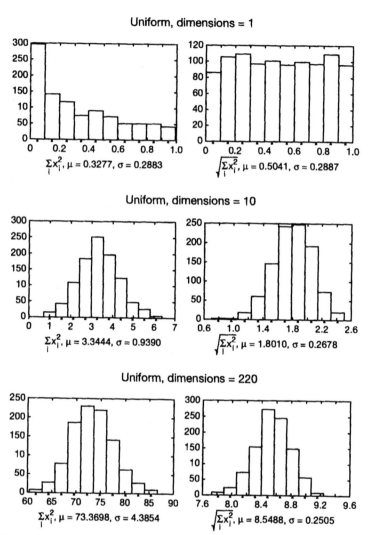

Figure 5-12. Histograms of functions of Uniformly distributed random variables.

The diagonals are nearly orthogonal to all coordinate axis [59, 60].

The cosine of the angle between any diagonal vector and a Euclidean coordinate axis is

$$\cos(\theta_d) = \pm \frac{1}{\sqrt{d}} \ ,$$

Figure 5-13 illustrates how the angle, θ_d, between the diagonal and the coordinates approaches $90°$ with increases in dimensionality. Note that $\lim_{d \to \infty} \cos(\theta_d) = 0$, which implies that in high-dimensional space the diagonals have a tendency to become

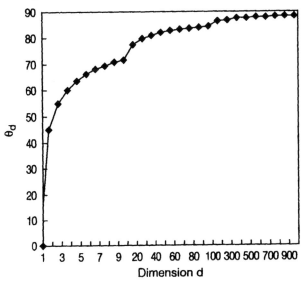

Figure 5-13. Angle (in degrees) between a diagonal and a Euclidean coordinate vs. dimensionality.

[59] D. W. Scott, *Multivariate density estimation*, Wiley, 1992, pp. 27-31.
[60] Wegman, E. J., Hyperdimensional data analysis using parallel coordinates, *Journal of the American Statistical Association*, Vol. 85, No. 411, pp. 664-675, 1990.

orthogonal to the Euclidean coordinates. This observation is important because the projection of any cluster onto any diagonal, e.g., by averaging features, could destroy information contained in multispectral data. In order to explain this, let a_{diag} be any diagonal in a d dimensional space. Let ac_i be the ith coordinate of that space. Any point in the space can be represented by the form:

$$P = \sum_{i=1}^{d} \alpha_i ac_i$$

The projection of P over a_{diag}, P_{diag} is:

$$P_{diag} = \left(P^T a_{diag}\right) a_{diag} = \sum_{i=1}^{d} \alpha_i \left(ac_i{}^T a_d\right) a_d$$

But as d increases $ac_i{}^T a_{diag} \approx 0$ which implies that $P_{diag} \approx 0$. As a consequence P_{diag} is being projected to the zero coordinate, losing information about its location in the d dimensional space.

The Required Number of Labeled Samples for Supervised Classification Increases as a Function of Dimensionality.

Fukunaga[61] proves that the required number of training samples is linearly related to the dimensionality for a linear classifier and to the square of the dimensionality for a quadratic classifier in an example circumstance. That fact is very relevant, especially since experiments have demonstrated that there are circumstances where second order statistics are more relevant than first order statistics in discriminating among classes in high dimensional data.[62] In terms of nonparametric classifiers the situation is even more severe. It has been estimated that as the number of dimensions increases, the

---

[61] K. Fukunaga, *Introduction to Statistical Pattern Recognition*, Academic Press, Inc., 1990.

[62] Chulhee Lee, and D. A. Landgrebe, Analyzing high dimensional multispectral data, *IEEE Transactions on Geoscience and Remote Sensing*, Vol. 31, No. 4, pp. 792-800, July, 1993.

sample size needs to increase exponentially in order to have an effective estimate of multivariate densities.[63, 64]

It is reasonable to expect that high-dimensional data contain more information so that more classes can be detected with greater accuracy. At the same time the characteristics above show that conventional techniques, which are usually based on computations at full dimensionality, may not deliver this advantage unless the available labeled training data is substantial. This was proved by Hughes[65] (see Chapter 3) who showed that with a limited number of training samples there is a penalty in classification accuracy as the number of features increases beyond some point.

For Most High-Dimensional Data Sets, Low-Dimensional Linear Projections Have a Tendency to be Normal, or a Combination of Normal Distributions, as the Dimension Increases.

It has been proved[66, 67] that as the dimensionality tends to infinity, lower dimensional linear projections will approach a normality model with probability approaching one (see Figure 5-14). Normality in this case implies a normal or a combination of normal

---

63   D. W. Scott, *Multivariate Density Estimation,* Wiley, 1992, pp. 208-212.

64   J. Hwang, S. Lay, A. Lippman, Nonparametric multivariate density estimation: A comparative study, *IEEE Transactions on Signal Processing,* Vol. 42, No. 10, 1994, pp. 2795-2810.

65   Hughes, G. F., On the mean accuracy of statistical pattern recognizers, *IEEE Transactions on Information Theory,* Vol. IT-14, No. 1, January 1968.

66   P. Diaconis, D. Freedman, Asymptotics of graphical projection pursuit. *The Annals of Statistics* Vol. 12, No 3 pp. 793-815, 1984

67   P. Hall, K. Li, On Almost Linearity Of Low Dimensional Projections From High Dimensional Data, *The Annals of Statistics,* Vol. 21, No. 2, pp. 867-889, 1993.

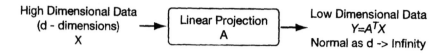

High Dimensional Data
(d - dimensions)
X
→ Linear Projection
A
→ Low Dimensional Data
$Y=A^TX$
Normal as d -> Infinity

Figure 5-14. The tendency of lower dimensional projections to be normal.

distributions. Several experiments illustrate this with simulated and real data. The procedure in these experiments was to project the data from a high dimensional space to a one-dimensional subspace. We will examine the behavior of the projected data as the number of dimensions in the original high dimensional space increases from 1 to 10 and finally to 100. The method of projecting the data is to multiply it with a normal vector with random angles from the coordinates. A histogram is used to observe the data distribution. A

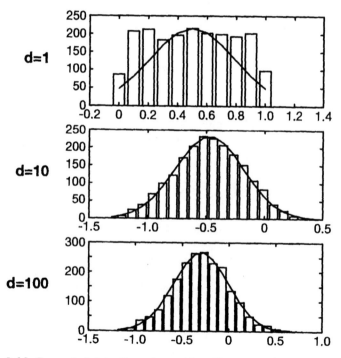

Figure 5-15. Generated data: One class with uniform distribution projected from a d dimensional space to one dimensional space.

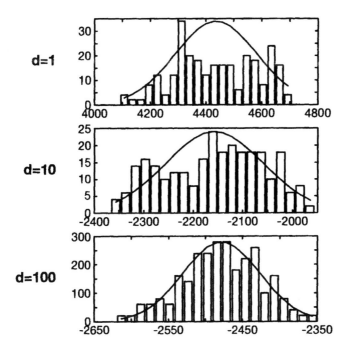

Figure 5-16. AVIRIS Multispectral data: One class, soybeans
in d dimensions projected to one dimension.

normal density function is plotted with the histogram to compare
the results to normal.

Figure 5-15 shows the case of generated data from a uniform
distribution. As the number of dimensions increases in the original
space, the histogram of the projected data has a tendency to be
normal. Figure 5-16 shows the results of the same experiment with
real AVIRIS data with one soybeans class. Note that the results are
similar to the generated data.

These results tempt us to expect that the data can be assumed to be
a combination of normal distributions in the projected subspace
without any problem. Other experiments show that a combination
of normal distributions where each one represents a different

261

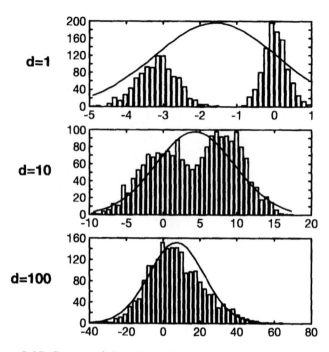

Figure 5-17. Generated data: Two classes with normal distributions
projected from d dimensions to one dimension.

statistical class could collapse into one normal distribution. That
would imply loss of information. Figures 5-17 and 5-18 show the
result of repeating the experiments for a two class problem. Both
figures illustrate the risk of damaging the information content, such
as separability of data when it is projected into one normal
distribution. In the case of Figure 5-18 we have real AVIRIS data
with a corn and a soybeans class.

In all these cases we can see the advantage of developing an
algorithm that will estimate the projection directions that separate
the explicitly defined classes, doing the computations in a lower-
dimensional space. The vectors that it computes will separate the
classes, and at the same time the explicitly defined classes will

262

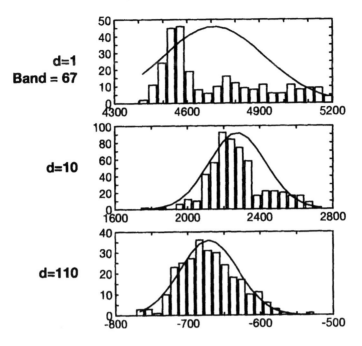

Figure 5-18. AVIRIS Multispectral data: Two classes, corn and
soybeans projected from d dimensions to one dimension.

behave asymptotically more like a normal distribution. The
assumption of normality will be better grounded in the projected
subspace than in full dimensionality.

5.5 Asymptotical First and Second Order Statistics Properties

An experimental result was shown in Figure 5-4 where some high-
dimensional data were classified in a manner to compare the relative
success of first and second order statistics. Here a more general
basis will be given for the use of the first- and second-order statistics
in hyperspectral data where adjacent bands can be correlated in
any way. The results will be based on the asymptotic behavior of
high-dimensional data. This will aid in understanding the conditions
required for the predominance of either first-order or second-order

statistics in the discrimination among the statistical classes in high-dimensional space.

It is reasonable to assume that as the number of features increases, the potential information content in multispectral data increases as well. In supervised classification that increment of information is translated to the number of classes and their separability. We will use Bhattacharyya distance here as the measure of separability. It provides a bound of classification accuracy taking into account first- and second-order statistics. Bhattacharyya distance is the sum of two components, one based primarily on mean differences and the other based on covariance differences.

The Bhattacharyya distance under the assumption of normality is computed by the equation:

$$B = \frac{1}{8}[\mu_f - \mu_g]^T\left[\frac{\Sigma_f + \Sigma_g}{2}\right]^{-1}[\mu_f - \mu_g] + \frac{1}{2}\ln\frac{\left|\frac{1}{2}[\Sigma_f + \Sigma_g]\right|}{\sqrt{|\Sigma_f||\Sigma_g|}}$$

We will denote B = Bhatt Mean + Bhatt Cov where Bhatt Mean, the first term on the right, is the mean difference component and Bhatt Cov, the second term on the right, is the covariance difference component.

In order to see how Bhattacharyya distance and its mean and covariance components can aid in the understanding of the role of first- and second-order statistics, two experiments are presented. The first one has conditions where second-order statistics are more relevant in discriminating among the classes. The second experiment has conditions for the predominance of first-order statistics.

Experiment 1

In this experiment data are generated for two classes. Both classes belong to normal distributions with different means and covariances. Each class has 500 points. Their respective parameters are

$$\mu_1 = \begin{bmatrix} 0 & 0 & 0 & 0 & 0 & 0 & 0 & 0 & 0 \end{bmatrix}^T$$

$$\mu_2 = \begin{bmatrix} 1.5 & 1 & .5 & 0 & 0 & 0 & 0 & 0 & 0 \end{bmatrix}^T$$

$$\Sigma_1 = \begin{bmatrix} 1 & & & & & & & 0 \\ & 1 & & & & & & \\ & & 1 & & & & & \\ & & & 1 & & & & \\ & & & & 1 & & & \\ & & & & & 1 & & \\ & & & & & & 1 & \\ 0 & & & & & & & 1 \end{bmatrix}, \quad \Sigma_2 = \begin{bmatrix} 1.5 & & & & & & & 0 \\ & 1 & 9 & & & & & \\ & & 3 & & & & & \\ & & & 3 & & & & \\ & & & & 3 & & & \\ & & & & & 3 & & \\ & & & & & & 3 & \\ 0 & & & & & & & 3 \end{bmatrix}$$

The data are classified by three classifiers, the ML classifier, the ML (ML Cov) classifier constrained to use only covariance difference, and the minimum distance classifier (Min Dist). This enables us to have similar conditions to the experiment of Figure 5-4. The results are shown in Figure 5-19.

The results resemble those of Figure 5-4. In order to better demonstrate the roles of first- and second-order statistics, the mean (Bhatt Mean) and covariance (Bhatt Cov) components of Bhattacharyya distance and their sum were computed as shown in

265

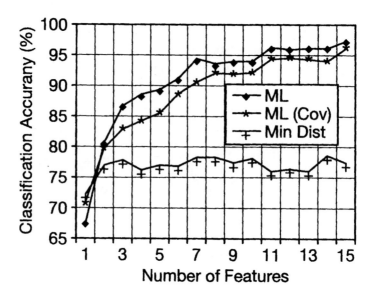

Figure 5-19. Performance comparison of Gaussian ML,
Gaussian ML with zero mean data, and Minimum Distance classifier. Two generated classes.

Figure 5-20. Their ratio of Bhatt Mean / Bhatt Cov was calculated
and shown in Figure 5-21.

Both figures show a relationship between second-order statistics
predominance and Bhatt Cov relevance. As the number of
dimensions increases, the ratio Bhatt Mean / Bhatt Cov decreases
significantly and the ML Cov classifier becomes more effective
than Min Dist. This shows that if, as the dimensionality increases,
the ratio Bhatt Mean / Bhatt Cov decreases, then second order
statistics are more relevant in high-dimensional data even when
that could not be the case in low-dimensionality.

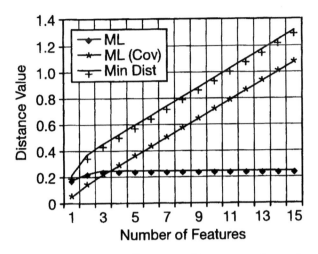

Figure 5-20. Bhattacharyya distance and its mean and covariance components.

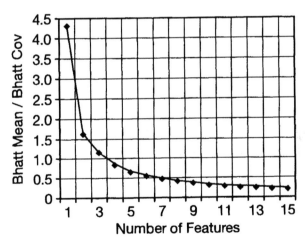

Figure 5-21. Ratio of Bhattacharyya distance mean component over the covariance component

Experiment 2

This experiment is similar to the previous one, with the difference that first-order statistics are predominant. The parameters of the two statistical classes are

$$\mu_1 = \begin{bmatrix} 0 & 0 & 0 & 0 & 0 & 0 & 0 & 0 & 0 & 0 \end{bmatrix}^T$$

$$\mu_2 = \begin{bmatrix} 1.5 & 1 & .5 & 0 & 0 & 0 & 0 & 0 & 0 & 0 \end{bmatrix}^T$$

$$\Sigma_1 = \begin{bmatrix} 1 & & & & & & & & & 0 \\ & 1 & & & & & & & & \\ & & 1 & & & & & & & \\ & & & 1 & & & & & & \\ & & & & 1 & & & & & \\ & & & & & 1 & & & & \\ & & & & & & 1 & & & \\ & & & & & & & 1 & & \\ & & & & & & & & 1 & \\ 0 & & & & & & & & & 1 \end{bmatrix} \qquad \Sigma_2 = \begin{bmatrix} 2.5 & & & & & & & & & 0 \\ & 2 & & & & & & & & \\ & & 1 & & & & & & & \\ & & & 1 & & & & & & \\ & & & & 1 & & & & & \\ & & & & & 1 & & & & \\ & & & & & & 1 & & & \\ & & & & & & & 1 & & \\ & & & & & & & & 1 & \\ 0 & & & & & & & & & 1 \end{bmatrix}$$

The classification results are shown in Figure 5-22. Observe that Min Dist classifier becomes more accurate than Min Cov after six dimensions. The mean (Bhatt Mean) and covariance (Bhatt Cov) components of Bhattacharyya distance and their sum were computed and are shown in Figure 5-23. Their ratio of Bhatt Cov / Bhatt Mean was calculated and shown in Figure 5-24. As the number of dimensions increases, the ratio Bhatt Cov / Bhatt Mean decreases showing that first-order statistics are more relevant in the classification of data.

These results indicate directly how the predominance of the mean or covariance Bhattacharyya distance components relates with first- or second-order statistics in terms of classification accuracy. Further

268

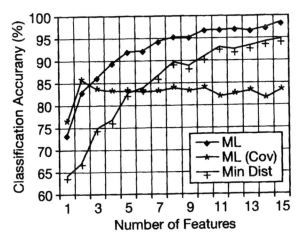

Figure 5-22. Performance comparison of Gaussian ML, Gaussian ML with zero mean data, and Minimum Distance classifier for two generated classes.

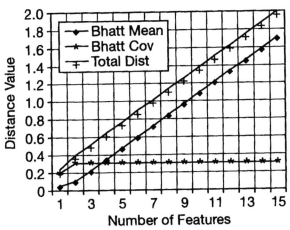

Figure 5-23. Bhattacharyya distance and its mean and covariance components.

269

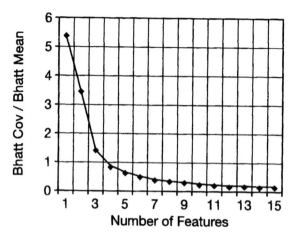

Figure 5-24. Ratio of Bhattacharyya distance covariance component over the mean component.

details about this division of separation are available in the primary reference[68] for this section.

5.6 High-Dimensional Implications For Supervised Classification

Based on the characteristics of high-dimensional data that the volumes of hypercubes have a tendency to concentrate in the corners, and for hyperellipsoids in outside shells, it should be apparent that high-dimensional space is mostly empty, and multivariate data are usually in a lower dimensional structure. As a consequence it is possible to reduce the dimensionality without losing significant information and separability. Because of the difficulties of density estimation in nonparametric approaches, a parametric version of data analysis algorithms may be expected to provide better performance where only limited numbers of labeled

---

[68] Luis Jimenez and David Landgrebe, "Supervised Classification in High Dimensional Space:Geometrical, Statistical and Asymptotical Properties of Multivariate Data," IEEE Transactions on Systems, Man, and Cybernetics, Volume 28 Part C Number 1, pp. 39-54, Feb. 1998.

samples are available to provide the needed a priori information. These observations make clear the importance of effective feature extractions processing in the analysis process. One should therefore find the optimal subset of features to use in each specific instance. Such procedures will be the subject of Chapter 6.

6 Feature Definition

In previous chapters the importance of feature spaces in the information extraction process was established, and that various special, case-specific feature spaces may be helpful. In this chapter a number of means for establishing such feature subspaces are described. The chapter begins with several methods for feature extraction that provide improved opportunities for operator visualization of spectral aspects. The chapter concludes with a description of means for extracting features optimal for classification purposes.

6.1 Introduction

Sensor systems are often designed to collect data in a larger number of spectral bands than would be needed for any specific application task, because each possible task is likely to be best served by a different set of features. However, using too many spectral features for an analysis has been shown to be detrimental to performance. Thus means are needed by which to determine the best set of features to use in any given classification circumstance.

Figure 6-1 shows how feature definition algorithms typically fit into the overall processing scheme. There can be a number of different objectives in determining the feature set to use in a specific instance. Some are useful for image enhancement tasks in order to make certain characteristics of a scene more apparent to a human observer. Others are intended for use in calculating optimal features in feature space for use with a pattern recognition algorithm. In the following, examples that fall into both of these categories are described.

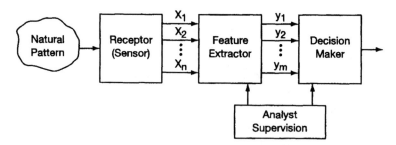

Figure 6-1. Feature extraction as a part of an overall information system.

6.2 Ad Hoc and Deterministic Methods

1. NDVI Ad hoc and deterministic methods are primarily of value for image enhancement purposes when the data are to be viewed or analyzed by manual image interpretation techniques. An example of one of the more common ones occurring in the literature is the Normalized Difference Vegetative Index (NDVI). It is a fairly unique characteristic of green vegetation that the reflectance in the near IR region is several times as large as that in a visible band, as seen in Figure 6-2. The idea is that the greater the Near-IR response over the visible, the larger is the vegetative index.

The general concept for NDVI is that it is to be calculated as

$$\text{NDVI} = \frac{\text{NIR} - \text{VIS}}{\text{NIR} + \text{VIS}}$$

where NIR is the digital count of a Near-IR band for a pixel, and VIS is the digital count for that pixel in a visible band. The sum in the denominator is used to normalize the quantity, thus the name. Note that this transformation is sensor dependent, since the value of NDVI would depend not only on the net sensor system gain in

274

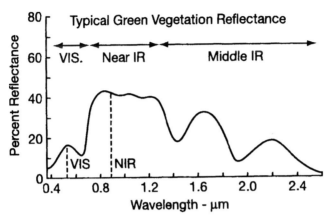

Figure 6-2. Percent reflectance vs. wavelength typical for green vegetation.

the channels used but on the specific bandwidth and location as well. Its value would also vary significantly depending on whether radiance data, which are affected by the solar radiance illumination curve, or reflectance data are used. NDVI thus provides a relative indication of the vegetation, rather than an absolute measure, and even there, it is dependent on the degree to which the relative size of the NIR band response to that in the visible implies vegetation and only vegetation. Its simple, straightforwardness has resulted in NDVI's wide use.

2. Tasseled Cap. The Kauth-Thomas Tasseled Cap Transformation[69] is another example of an ad hoc feature extraction transformation. Its purpose is again image enhancement for human view. It was introduced initially with regard to Landsat MSS data with a focus

---

[69] R. J. Kauth and G. S. Thomas, The Tasseled Cap - A graphic description of the spectral-temporal development of agricultural crops as seen by LANDSAT, *Proceedings of the LARS 1976 Symposium on Machine Processing of Remotely Sensed Data*, Purdue University.

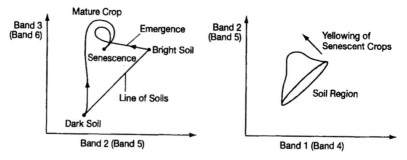

Figure 6-3. Characteristics of agriculture crops in Landsat MSS data.

on agricultural crops mapping problems. It was noted in this case that

- Soils tend to fall along a line in feature space in the band 5–band 6 plane.[70] See Figure 6-3.

- For a given crop, during canopy emergence, the temporal trajectory is toward a common green region implying closure of the canopy over the soil.

- From that time on, the yellowing of senescence occurs, taking the trajectory out of the plane defined by the green region and the line of soils.

In three dimensions the temporal trajectory for a crop appears as shown in Figure 6-4, thus the name, Tasseled Cap.

---

[70] The LANDSAT MSS bands are as follows (MSS bands were customarily numbered 4–7 instead of 1–4):

Band 1	0.5–0.6 μm	Green	(Band 4)
Band 2	0.6–0.7 μm	Red	(Band 5)
Band 3	0.7–0.8 μm	Reflective IR	(Band 6)
Band 4	0.8–0.1.1 μm	Reflective IR	(Band 7)

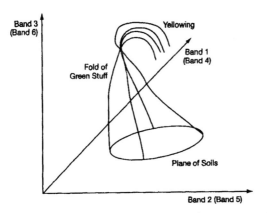

Figure 6-4. Three-dimensional view of the tasseled cap temporal trajectory.

This trajectory behavior led Kauth and Thomas to define a new set of coordinates as follows:

- First dimension: Defined along the soil line. This was referred to as the "soil brightness" direction.

- Second dimension: Orthogonal to the first and in the direction toward the green point, indicating the greenness.

- Third dimension: Orthogonal to the first two and in the yellowing direction.

- The fourth dimension: Determined to account for the remaining spectral variation. It was referred to as "non-such."

The resulting transformation (valid only for MSS, and in fact only for LANDSAT 2 MSS) is

$$\mathbf{u} = \mathbf{Rx} \quad \text{where } \mathbf{R} = \begin{bmatrix} 0.433 & 0.632 & 0.586 & 0.264 \\ -0.290 & -0.562 & 0.600 & 0.491 \\ -0.829 & 0.522 & -0.039 & 0.194 \\ 0.223 & 0.012 & -0.543 & 0.810 \end{bmatrix}$$

Note from this matrix that the Tasseled Cap Transformation is as follows:

- Soil Brightness (first line of matrix) is the weighted sum of the four bands with approximately equal emphasis.

- Greeness (second line) is largely the difference between the visible and IR responses.

- Yellowness is substantially the difference between the responses in the visible red and green bands.

Such a transformation, while valid only for the sensor system for which it is derived, can be useful in providing a feature space in which human perception of what the interpretation is of a given pixel response is enhanced. It does not enhance separability, however, since a linear transformation at most rotates and translates the coordinate system, but it cannot change the relative separability between two distributions.

6.3 Feature Selection

Perhaps the most straightforward way to reduce the dimensionality of a data set is to simply select a subset of spectral bands on some optimal basis. Such a task was illustrated in chapter 3 using the Divergence statistical separability measure. For completeness, an example will be repeated here, only this time using the

Bhattacharrya distance discussed earlier. Suppose, for example, one wishes to determine the best four bands out of seven available to use in classifying a given data set for which five classes have been defined by training samples. To do this one must compute the Bhattacharrya distance between each pair of classes for each subset of size four from the seven-band data. Next, arrange the results in a table as shown in Table 6-1.

class pair symbols>		12	13	14	15	23	24	25	34	35	45
Channels	Ave.	Interclass Bhattacharrya Distances									
1. 2 5 6 7	8.05	1.72	14.6	4.53	16.2	5.89	1.28	8.16	1.27	20.6	6.11
2. 1 2 5 7	8.00	1.12	15.5	4.75	16.2	6.49	1.59	6.83	1.09	21.7	4.54
3. 3 5 6 7	7.91	1.80	15.8	4.57	15.4	6.85	1.41	7.34	1.28	18.3	6.14
4. 2 3 5 7	7.85	1.39	15.7	4.22	17.7	6.69	1.46	6.95	1.18	19.0	4.08
5. 2 4 5 7	7.73	1.68	14.6	3.93	18.4	6.00	1.27	6.64	1.56	19.0	4.20
6. 3 4 5 7	7.53	1.78	15.9	4.18	15.5	7.30	1.53	6.09	1.56	16.9	4.34
7. 1 3 5 7	7.47	1.40	16.1	4.81	15.4	6.80	1.66	6.06	1.15	17.1	4.10
8. 1 5 6 7	7.31	1.71	15.4	5.06	11.9	6.64	1.55	6.01	1.21	18.2	5.22
9. 1 4 5 7	7.12	1.70	15.4	4.72	13.9	6.96	1.62	4.49	1.50	17.1	3.70
10. 4 5 6 7	7.03	1.81	13.8	4.34	12.1	6.20	1.21	5.18	1.62	18.6	5.27

Table 6-1. A table of Bhattacharyya distances between pairs of classes.

In Table 6-1 the interclass pair distances for each possible 4-tuple appear in rows placed in descending order based on the average distance in the row. Only the top 10 of the 35 possible rows are shown. The results for this example suggest then that the 4-tuple of bands 2, 5, 6, and 7 would be good to use for classifying the five classes involved.

However, notice that the discriminations between classes 1 and 2 and also that between 2 and 4 and 3 and 4 are particularly challenging. Row 9 with channels 1, 4, 5, and 7 might be a better choice in terms of the expected overall accuracy, since these features seem to provide better performance on the difficult classes without sacrificing much on the easier classes.

In addition to the need for skill in subjective judgments of the type just illustrated, this approach to feature selection has two perhaps significant limitations. For one, the method is usable only in relatively low dimensional cases. As the number of bands available increases, the calculations involve become unwieldy. To select an m feature subset out of a total of n features, the number N of band combinations (rows) that must be calculated is given by the binomial coefficient

$$N = \binom{n}{m} = \frac{n!}{(n-m)!\,m!}$$

This number can become quite large. For example, to find the best 10 features out of 20, there are 184,756 combinations that must be evaluated. For the best 10 out of 100, the number of combinations exceeds 1.7×10^{13}. To mitigate this computational load problem, one might use a sequential forward (choose the best 1 out of 100, then choose the best one to add out of the remaining 99, etc.) or sequential backward search. While workable, these are suboptimal compared to the exhaustive search suggested in the example above.

The scheme in the example above is suboptimal in another sense, its exclusionary nature. Choosing the best 4 of 7 features completely excludes any separation characteristics that the excluded 3 might have. Even though the excluded features may have less useful separability characteristics, they may not be zero. This also becomes especially significant for the higher dimensional hyperspectral case, where one might desire the best 10 of 100 features. The excluded 90 features might contain a significant amount of useful characteristics. It is because of these two limitations in the hyperspectral case that the feature extraction methods described below have become of special utility.

6.4 Principal Components / Karhunen-Loeve

Suppose that one has data distributed over a two-dimensional region as shown in Figure 6-5.

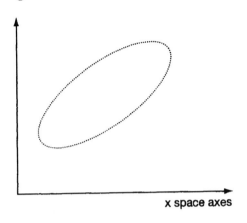

Figure 6-5. Hypothetical distribution of data in 2-dimensional feature space.

Since the region is diagonally oriented, data from the two axes are correlated. Define a new set of axes as in Figure 6-6, such that the data are horizontally or vertically oriented. In this new space, the

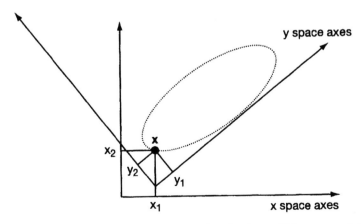

Figure 6-6. Rotation of the coordinates so that the data will be uncorrelated.

281

data will not be correlated, and the y_1 component will have a greater dynamic range than y_2 or either of the x components, thus the "principal" component name.

To find the necessary transformation from **x** to **y**, find the mean vector

$$\hat{\boldsymbol{\mu}}_x = E[\mathbf{x}] = \frac{1}{K}\sum_{i=1}^{K}\mathbf{x}_i$$

(assuming K samples) and the covariance matrix of the data set,

$$\hat{\Sigma}_x = \frac{1}{K-1}\sum_{i=1}^{K}(\mathbf{x}_i - \hat{\boldsymbol{\mu}})(\mathbf{x}_i - \hat{\boldsymbol{\mu}})^{T}$$

Now the linear transformation $\mathbf{y} = \mathbf{A}\mathbf{x}$ is desired such that Σ_y, the covariance matrix in the transformed coordinate system, will be diagonal (i.e., have all elements off the diagonal equal zero). This means that all features are uncorrelated with one another. It is well known that the transformation which accomplishes this is the one which satisfies the equation, $\Sigma_y = \mathbf{A}\Sigma_x\mathbf{A}^{T}$, where Σ_y has the form

$$\Sigma_y = \begin{bmatrix} \lambda_1 & 0 & \cdots & 0 \\ 0 & \lambda_2 & & \\ \vdots & & \ddots & \\ 0 & & & \lambda_N \end{bmatrix}$$

and is found by solving the equation $|\Sigma_x - \lambda\mathbf{I}| = 0$. The transformation **A** is referred to as the principal components transformation, and the continuous variable version is known as the Karhunen-Loeve transformation. The solution to this eigenvalue equation results in an Nth order polynomial equation that can be solved for the N eigenvalues, $\{\lambda_n\}$. Once obtained, the eigenvalues are arranged in descending order, and substituted into the equation $[\Sigma_x - \lambda_n I] \, a_n = 0$ to obtain the n transformation vectors, $\{a_n\}$. In doing so, use is made of the fact that the transformation vectors are to be normalized, meaning that the sum of their squares must be one.

Note the effect of placing the eigenvalues, i. e., the elements on the diagonal of Σ_y, in descending order. Since the eigenvalues are now the variance of the variables in the new space, the direction of the y_1 axis is the direction of maximum variance, that of y_2 is orthogonal to y_1 in the direction of maximum remaining variance, and so on. Thus, this transformation provides a convenient means for ordering the features and selecting those that will make the greatest contribution to representation of the data (but not necessarily to classification!). This is the basis for using this transformation for feature extraction.

Following is a brief example illustrating the effect of a principal component transformation. Shown in Figure 6-7 are images of the first four Thematic Mapper bands for a representative agricultural area. It is seen that there is good contrast in all four bands indicating a moderated dynamic range and thus a normal variance for the data in each band. The actual variances for the four bands are 50.41, 24.01, 75.69, and 408.04, respectively.

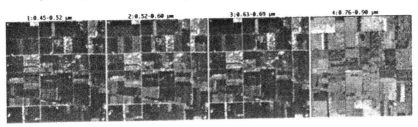

Figure 6-7. Four band Thematic Mapper data in image space.

A principal component calculation was carried out on this data with the result shown below. The eigenvalues for the new features are the variances of the data in the new coordinates. It is seen that the first principal component now has a variance very much larger than the others, and only the first two have a variance of significant size.

Component	Eigenvalue	Percent	Cum. Percent
1	460.44	82.32	82.32
2	93.75	16.76	99.09
3	3.15	0.56	99.65
4	1.93	0.35	100.00

The eigenvector coefficients are given in the following table.

Components	Channel 1	2	3	4
1	0.2124	0.1242	0.2807	-0.9277
2	0.5451	0.4052	0.6334	0.3707
3	-0.8107	0.2829	0.5125	0.0073
4	0.0208	-0.8604	0.5073	0.0431

Thus the new, transformed features y_{jn} would be created from the original data values x_{mn} as

$$y_{1n} = 0.21241x_{1n} + 0.12417x_{2n} + 0.28065x_{3n} - 0.92774x_{4n}$$
$$y_{2n} = 0.54514x_{1n} + 0.40520x_{2n} + 0.63343x_{3n} + 0.37067x_{4n}$$

$$\cdots \cdots$$

where y_{jn} is the jth component of the nth pixel.

Images made with the new components are shown in Figure 6-8. It is seen that the first two have significant contrast, but the dynamic range of the last two is so small as to show mostly noise. Indeed, the contrast in feature 1, as quantified by the eigenvalue of feature 1, is now so large that it is greater than can be seen in an image display of the data. Note that the contrast in feature image 1 does not appear too different from that in feature image 2, even though the ratio of the two eigenvalues is more than 4:1.

Figure 6-8. Images of the data of Figure 6-7 after a principal components transformation.

This change can be seen more clearly by viewing variance comparisons and the histograms of the original data and comparing it with that of the transformed data, as shown in Figure 6-9. Notice in Figure 6-9 that while the original histograms in all seven bands

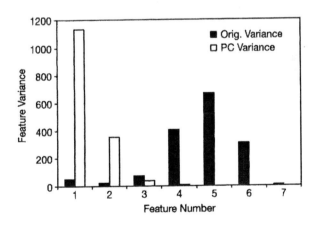

Figure 6-9. Variance comparisons for before and after a principal components transformation.

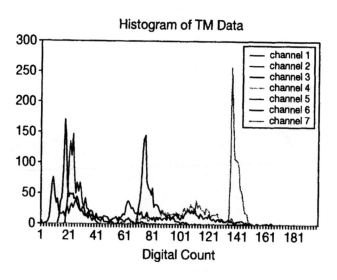

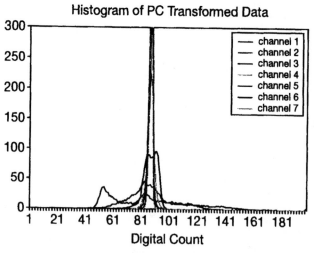

Figure 6-9 (continued). Histograms comparisons for before and after a
principal components transformation. (In color on CD)

have reasonably wide dynamic ranges, after the transformation,
only two have very much dynamic range breadth. The last five
components have all of their pixels with approximately the same
digital count. This same characteristic is displayed in the variance

comparison, where only the first two components have variance of significant size.

6.5 Discriminant Analysis Feature Extraction (DAFE)

The principal components transformation is based on the global covariance matrix of the entire data set and is thus not sensitive explicitly to inter-class structure. It often works as a feature reduction tool because, in remote sensing data, classes are frequently distributed in the direction of maximum data scatter. Discriminant Analysis Feature Extraction is a method to enhance separability. Consider the hypothetical situation shown in Figure 6-10, where the overall data distribution indicated in Figure 6-5 is now shown in terms of the area of concentration of its two constituent classes.

This simple 2-dimensional hypothetical situation will be used to illustrate an approach for reducing the dimensionality from two to one while enhancing the class separability.

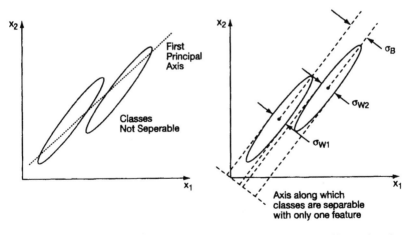

Figure 6-10. A hypothetical distribution of two classes in 2-dimensional feature space.

- On the left drawing it is seen that the classes are not separable with one feature in the original space nor with regard to the principal component axes.

- On the right it is seen that the classes would be separable with one feature if the dotted axis could be found. The problem is to find this axis.

Inspection of the figure reveals that the primary axis of this new transformation should be oriented such that the classes have the maximum separation between their means on this new axis, while at the same time they should appear as small as possible in their individual spreads. If the former is characterized by σ_B in the figure and the latter by σ_{w1} and σ_{w2}, then it is desired to find the new axis such that

$$\frac{\sigma_B^2}{\sigma_W^2} = \frac{\text{between} - \text{class variance}}{\text{average within} - \text{class variance}}$$

is maximized. Designate Σ_W as the average of σ_{w1} and σ_{w2}. In matrix form the within-class scatter matrix Σ_W and the between-class scatter matrix Σ_B may be defined as,[71]

$$\Sigma_W = \sum_i P(\omega_i)\Sigma_i \qquad \text{(within class scatter matrix)}$$

$$\Sigma_B = \sum_i P(\omega_i)(\mu_i - \mu_o)(\mu_i - \mu_o)^T$$

(between class scatter matrix)

$$\mu_o = \sum_i P(\omega_i)\mu_i$$

[71] K. Fukunaga, *Introduction to Statistical Pattern Recognition,* Academic Press, 1972.

Here μ_i, Σ_i, and $P(\omega_i)$ are the mean vector, the covariance matrix, and the prior probability of class ω_i, respectively. The criterion for optimization may be defined as

$$J_{DAFE} = tr(\Sigma_W^{-1} \Sigma_B)$$

New feature vectors are selected to maximize the criterion. This is the basis for Discriminant Analysis Feature Extraction (DAFE).

However, this method does have some shortcomings too. For example, since discriminant analysis mainly utilizes class mean differences, the feature vector selected by discriminant analysis is not reliable if mean vectors are near to one another. And, by using the lumped covariance in the criterion, discriminant analysis may lose information contained in class covariance differences. Another problem with the criteria function using scatter matrices is that the criteria generally do not have a direct relationship to the error probability. Perhaps more significantly, the method only produces optimal features up to one less than the number of classes.

6.6 Decision Boundary Feature Extraction (DBFE)

The use of pattern recognition in the field of remote sensing has some special characteristics not found in a number of other fields. In particular, in order to minimize the effects of uncontrollable observations variables, one must retrain the classifier for each data set to be analyzed. This enhances the importance of using a proper training process and the choice of features to be used. Further, it is usually the case that there is a paucity of training samples, and increasingly as time goes on, the number of spectral bands (features) available is increasing. This causes special concern with regard to class statistics estimation error as illustrated by the Hughes phenomena. These characteristics place a premium upon finding both the *best features* of a given dimensionality, and the *best dimensionality* to use.

An alternate approach to optimal feature design may be taken to determine a procedure that best meets these requirements.[72] The approach uses the training samples directly, instead of statistics derived from them. The training samples are used to determine the location of the decision boundary implied by them, and from this to define a transformation that results in optimal features, as the following development will show.

The Bayes Decision Rule may be restated as

$$p(\omega_1)p(\mathbf{X}|\omega_1) > p(\omega_2)p(\mathbf{X}|\omega_2) \sim \mathbf{X} \in \omega_1$$
$$p(\omega_1)p(\mathbf{X}|\omega_1) < p(\omega_2)p(\mathbf{X}|\omega_2) \sim \mathbf{X} \in \omega_2$$

Let $\qquad h(\mathbf{X}) = -\ln \{p(\mathbf{X}|\omega_1)\} + \ln \{p(\mathbf{X}|\omega_2)\}$
and,

$$t = \ln\left\{\frac{p(\omega_1)}{p(\omega_2)}\right\}$$

Then the Bayes decision rule becomes
$$h(\mathbf{X}) < t, \sim \mathbf{X} \in \omega_1$$
$$h(\mathbf{X}) > t, \sim \mathbf{X} \in \omega_2$$

Discriminantly Redundant Features

Let $\{\beta_1, \beta_2, \dots, \beta_N\}$ be the basis of a possible new feature space representation. Then, for an N band measure of pixel response,

---

[72] Chulhee Lee and D. A. Landgrebe, Feature extraction based on decision boundaries, *IEEE Transactions on Pattern Analysis and Machine Intelligence*, Vol. 15, No. 4, pp. 388-400, April 1993.

$$\mathbf{x} = \sum_{i=1}^{N} a_i\beta_i \quad \text{and} \quad \hat{\mathbf{x}} = \sum_{\substack{i=1 \\ i \neq k}}^{N} a_i\beta_i$$

that is, $\hat{\mathbf{x}}$ approximates. $\mathbf{x}$ in original space but without β_k.

- Definition: We say β_k is discriminantly redundant if for all observations $\mathbf{x}$

$$\{h(\mathbf{x}) - t\} \ \{h(\hat{\mathbf{x}}) - t\} > 0$$

that is, with or without β_k, the class remains the same.

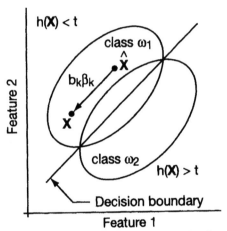

Figure 6-11. An example of a discriminantly redundant feature, β_k.

Discriminantly Informative Features
- Definition: We say that β_k is discriminantly informative if there exists at least one observation $\mathbf{x}$ such that,

$$\{h(\mathbf{x}) - t\} \ \{h(\hat{\mathbf{x}}) - t\} < 0$$

that is, without β_k, the classification changes.

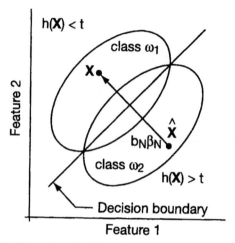

Figure 6-12. Example of a discriminantly informative feature, β_i.

Effective Decision Boundary

We take the definition of the decision boundary as $\{x \mid h(x) = t\}$. It can be a point, line, surface, or hypersurface. Although it may extend to infinity, in most cases some portion of it is not significant in discriminating between classes. Thus for practical purposes, we define the EFFECTIVE DECISION BOUNDARY as

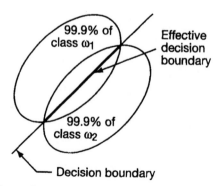

Figure 6-13. Example. $\mathbf{M}_1 \neq \mathbf{M}_2$, $\mathbf{\Sigma}_1 = \mathbf{\Sigma}_2$. The decision boundary is a straight line, and the effective decision boundary is a line segment coincident with it.

292

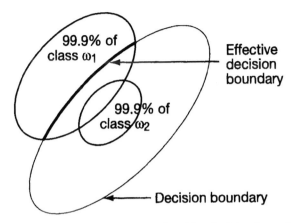

Figure 6-14. Example. $M_1 \neq M_2$, $\Sigma_1 \neq \Sigma_2$. The decision boundary and the
effective decision boundary.

$$\{x \mid h(x) = t, x \in R_1 \text{ or } x \in R_2\}$$

where R_1 is the smallest region that contains a certain portion, $P_{threshold}$, of class ω_1 and R_2 is the smallest region that contains a certain portion, $P_{threshold}$, of class ω_2. Thus the effective decision boundary may be seen as the intersection of the decision boundary and the regions where most of the data are located. Figures 6-13 and 6-14 show some examples, with $P_{threshold}$ set to 99.9%.

Intrinsic Discriminant Dimension

- Definition: The Intrinsic discriminant dimension is the smallest dimension M of a subspace W such that for any observation x

$$\{h(x) - t\}\{h(\hat{x}) - t\} > 0$$

Here $\hat{x} = \sum_{i=1}^{M} a_i \phi_i$, is an approximation of x in a transformed space with basis

$$\{\phi_1, \phi_2, \dots, \phi_M\}, \quad M \leq N$$

293

The intrinsic discriminant dimension is thus seen to be the smallest dimensional subspace wherein the same classification accuracy can be obtained as could be obtained in the original space.

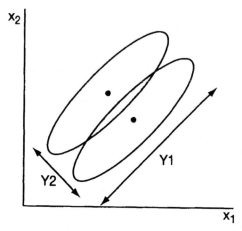

Figure 6-15. An example where the Intrinsic Discriminant Dimension = 1

Redundancy Testing.

- Lemma: If vector V is orthogonal to the vector normal to the decision boundary at every point on the decision boundary, vector V contains no information useful in discriminating between classes, i.e., vector V is discriminantly redundant.

- Lemma: If vector **V** is normal to the decision boundary at at least one point on the decision boundary, then vector **V** contains information useful in discriminating between classes, i.e., vector **V** is discriminantly informative.

Decision Boundary Feature Matrix

It can be seen that a vector normal to the decision boundary at a point is a discriminantly informative feature, and the effectiveness of the vector is roughly proportional to the area of the decision boundary that has the same normal vector. Now we can define a DECISION BOUNDARY FEATURE MATRIX that we can use to predict the intrinsic discriminant dimension and find the necessary feature vectors.

Let $N(x)$ be the unit normal vector to the decision boundary at a point x on the decision boundary. Then the DECISION BOUNDARY FEATURE MATRIX is defined as

$$\Sigma_{DBFM} = \frac{1}{K} \int_S N(x)N^t(x)p(x)dx$$

where $p(x)$ is a probability density function, $K = \int_S p(x)dx$, S is the decision boundary, and the integral is performed over the decision boundary. Some simple examples will help to make the concept clearer.

Example 1. Equal Covariances

Let

$$\mu_1 = \begin{vmatrix} 1 \\ 2 \end{vmatrix} \qquad \mu_2 = \begin{vmatrix} 2 \\ 1 \end{vmatrix} \qquad \Sigma_1 = \begin{vmatrix} 1 & 0.5 \\ 0.5 & 1 \end{vmatrix} = \Sigma_2$$

And $P(\omega_1) = P(\omega_2) = 0.5$ These distributions are shown in the Figure 6-16 as interlocking "ellipses of concentration." In a two-class, two-dimensional pattern classification problem, if the covariance matrices are the same, the decision boundary will be a straight line and the intrinsic discriminant dimension will be one. This means that the vector normal to the decision boundary at any

295

point is the same. The decision boundary feature matrix will be given by

$$\Sigma_{DBFM} = \frac{1}{K}\int_S N(\mathbf{x})N^t(\mathbf{x})p(\mathbf{x})d\mathbf{x} = \frac{1}{K}NN^t\int_S p(\mathbf{x})d\mathbf{x} = NN^t$$

$$\Sigma_{DBFM} = \frac{1}{\sqrt{2}}[-1,1]^t\frac{1}{\sqrt{2}}[-1,1] = \frac{1}{2}\begin{vmatrix} 1 & -1 \\ -1 & 1 \end{vmatrix}$$

$$Rank(\Sigma_{DBFM}) = 1$$

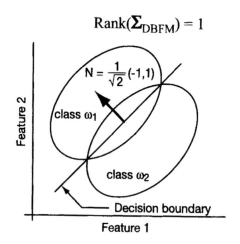

Figure 6-16. Example showing the circumstances of the decision boundary for Example 1.

Note that the rank of the decision boundary feature matrix is one, which is equal to the intrinsic discriminant dimension, and that the eigenvector corresponding to the nonzero eigenvalue is the desired feature vector, which gives the same classification accuracy as in the original 2-dimensional space.

Example 2. Equal Means.

Let

$$\mu_1 = \begin{vmatrix} 5 \\ 5 \end{vmatrix} \quad \Sigma_1 = \begin{vmatrix} 3 & 0 \\ 0 & 3 \end{vmatrix} \quad \mu_2 = \begin{vmatrix} 5 \\ 5 \end{vmatrix} \quad \Sigma_2 = \begin{vmatrix} 1 & 0 \\ 0 & 1 \end{vmatrix}$$

and $P(\omega_1) = P(\omega_2) = 0.5$

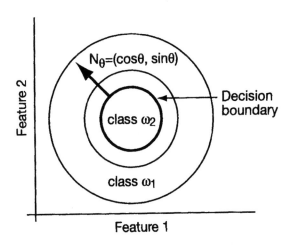

Figure 6-17. The circumstances of the decision boundary for Example 2.

The distributions of the two classes are shown in Figure 6-17 as concentric ellipses of concentration. In this example, the decision boundary is a circle and symmetric, and $\frac{1}{K} p(\mathbf{X})$ is a constant given by $\frac{1}{2\pi r}$ where r is the radius of the circle. The decision boundary feature matrix is given by

$$\Sigma_{DBFM} = \int_0^{2\pi} \frac{1}{2\pi} [\cos\theta \sin\theta]^t [\cos\theta \sin\theta] r \, d\theta$$

297

$$= \frac{1}{2\pi} \int_0^{2\pi} \begin{bmatrix} \cos\theta\cos\theta & \cos\theta\sin\theta \\ \sin\theta\cos\theta & \sin\theta\sin\theta \end{bmatrix} d\theta$$

$$= \frac{1}{2\pi} \begin{bmatrix} \pi & 0 \\ 0 & \pi \end{bmatrix} = \frac{1}{2} \begin{bmatrix} 1 & 0 \\ 0 & 1 \end{bmatrix}$$

The Rank(Σ_{DBFM}) = 2

From the distribution of the data in Figure 6-17, it is seen that two features are needed to achieve the same classification accuracy as in the original space. This means that the intrinsic discriminant dimension is 2 in this case. Notice that the rank of the decision boundary feature matrix is also 2.

The EFFECTIVE DECISION BOUNDARY FEATURE MATRIX (EDBFM) can be defined in an identical manner as the DBFM except that S becomes the effective decision boundary. We note in passing that the DBFM of the whole boundary can be expressed as a summation of the DBFM's calculated from segments of the decision boundary if the segments are mutually exclusive and exhaustive. This property is useful in extending the utility of the DBFM to the multiclass case.

Intrinsic Discriminant Dimension Theorem
Using the definitions above, it is possible to prove the following:

- Theorem. The rank of the DECISION BOUNDARY FEATURE MATRIX (DBFM) of a pattern recognition problem is equal to the intrinsic discriminant dimension.

- Lemma. The eigenvectors of the DBFM corresponding to nonzero eigenvalues are the necessary feature vectors to achieve the same classification accuracy as in the original space.

Feature Extraction Procedure Based on the Effective Decision Boundary

The following procedure for the two-class case may be used to determine the transformation needed to find the desired minimal set of features.

1. Let $\hat{\mu}_i$ and $\hat{\Sigma}_i$ be the estimated mean and covariance of class ω_i. Classify the training samples using full dimensionality. Apply a chi-square threshold test to the correctly classified training samples of each class and delete outliers. In other words, for class ω_i, retain x only if

$$(x - \hat{\mu}_i)^t \hat{\Sigma}_i^{-1} (x - \hat{\mu}_i) < R_{t1}$$

In the following steps, only correctly classified training samples that passed the chi-square threshold test will be used. Let $\{x_1, x_2, ..., x_{L_1}\}$ be such training samples of class ω_1 and $\{y_1, y_2, ..., y_{L_2}\}$ be such training samples of class ω_2.

2. Apply a chi-square threshold test of class ω_1 to the samples of class ω_2 and retain y_j only if

$$(y_j - \hat{\mu}_i)^t \hat{\Sigma}_i^{-1} (y_j - \hat{\mu}_i) < R_{t2}$$

If the number of the samples of class ω_2 which pass the chi-square threshold test is less than L_{min} (see below), retain the L_{min} samples of class ω_2 that gives the smallest values.

299

3. For x_i of class ω_1, find the nearest sample of class ω_2 retained in step 2.
4. Find the point P_i where the straight line connecting the pair of samples found in step 3 meets the decision boundary.
5. Find the unit normal vector, N_i, to the decision boundary at the point P_i found in step 4. This may be found by,[73]

$$N = \nabla h(x)\big|_{x=x_0} = \left(\hat{\Sigma}_1^{-1} - \hat{\Sigma}_2^{-1} \right)x_0 + \left(\hat{\Sigma}_2^{-1}\mu_1 - \hat{\Sigma}_1^{-1}\mu_2 \right)$$

6. By repeating steps 3 through 5 for x_i, $i=1,..,L_1$, L_1 unit normal vectors will be calculated. From the normal vectors, calculate an estimate of the effective decision boundary feature matrix (Σ) from class ω_1 as follows:

$$\Sigma_{EDBFM}^1 = \frac{1}{L_1} \sum_i^{L_1} N_i N_i^t$$

Repeat steps 2 through 6 for class ω_2.
7. Calculate an estimate of the final effective decision boundary feature matrix as follows:

$$\Sigma_{EDBFM} = \frac{1}{2} \left(\Sigma_{EDBFM}^1 + \Sigma_{EDBFM}^2 \right)$$

The chi-square threshold test in step 1 is necessary to eliminate outliers. Otherwise, outliers may give a false decision boundary when classes are well separable. The chi-square threshold test for the other class in step 2 helps determine an effective decision boundary. Otherwise, the decision boundary feature matrix may be

---

[73] The derivation for this equation may be found in Appendix of Chulhee Lee and D. A. Landgrebe, Feature extraction and classification algorithms for high dimensional data, PhD thesis, Purdue University, December 1992, and School of Electrical Engineering Technical Report TR-EE 93-1, January 1993.

calculated from an insignificant portion of the decision boundary, resulting in ineffective features. In the experiments below, L_{min} in step 2 is set to 5 and R_{t1} is chosen such that

$$Pr\{(\mathbf{x} - \hat{\mathbf{\mu}}_i)^t \hat{\Sigma}_i^{-1}(\mathbf{x} - \hat{\mathbf{\mu}}_i) < R_{t1}\} = 0.95, i = 1, 2 \text{ and } R_{t1} = R_{t2}$$

The threshold probability is taken as 0.95. In the ideal case assuming a Gaussian distribution, the threshold probability can be larger, e.g., 0.999. However, for real data, if the threshold probability is set too large, some outliers could be included, causing reduced effectiveness in the resulting decision boundary feature matrix. Figure 6-18 shows an illustration of the procedure.

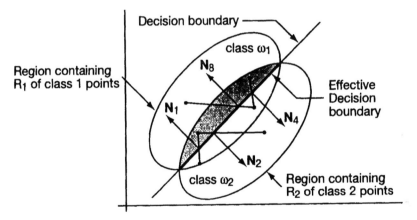

Figure 6-18. Illustration of the feature extraction procedure based on the effective decision boundary

For the multiclass case,

$$\Sigma_{EDBFM} = \sum_{i}^{M} \sum_{j, j \neq i}^{M} P(\omega_i)P(\omega_j)\Sigma_{E\,DBFM}^{ij}$$

where Σ^{ij}_{EDBFM} is the decision boundary feature matrix between class ω_i and class ω_j and $P(\omega_i)$ is the prior probability of class ω_i. In order to test the algorithm and to explore some of its characteristics, consider the following examples.

Example 1 (Generated Data).
(Gaussian Distribution, Same Covariance). Generate 300 samples for each of two classes of Gaussianly distributed data where,

$$\mu_1 = \begin{bmatrix} -1 \\ 1 \end{bmatrix} \quad \mu_2 = \begin{bmatrix} 1 \\ -1 \end{bmatrix} \quad \Sigma_1 = \begin{bmatrix} 1 & 0.5 \\ 0.5 & 1 \end{bmatrix} = \Sigma_2$$

$$P(\omega_1) = P(\omega_2) = 0.5$$

The eigenvalues, λ_i, and the eigenvectors, ϕ_i, of Σ_{EDBFM} which result are as follows.

$$\lambda_1 = 0.99995, \quad \lambda_2 = 0.00005, \quad \phi_1 = \begin{vmatrix} 0.7 \\ -0.71 \end{vmatrix} \quad \phi_2 = \begin{vmatrix} 0.71 \\ 0.70 \end{vmatrix}$$

From the eigenvalues it is seen that $\text{Rank}(\Sigma_{EDBFM}) = 1$. The distribution of the classes and the resulting decision boundary are shown in Figure 6-19.

Experiment 2 (Generated Data)
(Gaussian Distribution, No Mean Difference). Generate 300 samples for each of two classes of Gaussianly distributed data where

$$\mu_1 = \begin{bmatrix} 0.01 \\ 0 \end{bmatrix} \quad \mu_2 = \begin{bmatrix} -0.01 \\ 0 \end{bmatrix} \quad \Sigma_1 = \begin{bmatrix} 3 & 0 \\ 0 & 3 \end{bmatrix} \quad \Sigma_2 = \begin{bmatrix} 3 & 0 \\ 0 & 1 \end{bmatrix}$$

$$P(\omega_1) = P(\omega_2) = 0.5$$

The eigenvalues, λ_i, and the eigenvectors, ϕ_i, of Σ_{DBFM} which result are as follows.

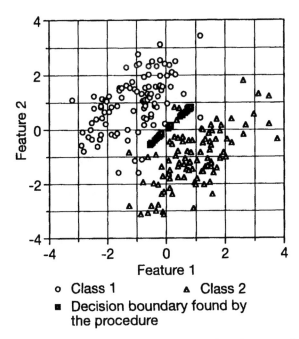

Figure 6-19. Samples and decision boundary for example 1.

$$\lambda_1 = 0.92421, \quad \lambda_2 = 0.07579, \quad \phi_1 = \begin{bmatrix} 0.06 \\ 1.00 \end{bmatrix} \quad \phi_2 = \begin{bmatrix} -1.00 \\ 0.06 \end{bmatrix}$$

Given that the two classes have different covariances, the decision boundary will be a second-order curve or curves and, in fact, in this case will be hyperbolas. Thus Rank(Σ_{EDBFM}) = 2, and two features are needed to achieve the same accuracy as in the original space. However, from the eigenvalues it is seen that, though not zero, the size of λ_2 is very much smaller than λ_1. Therefore nearly the same accuracy should be obtainable with one feature as with two. This turns out to be the case, as 61.0% accuracy is obtained using either $\{\phi_1\}$ and $\{\phi_1, \phi_2\}$. The distribution of the classes and the resulting decision boundary are shown in Figure 6-20.

303

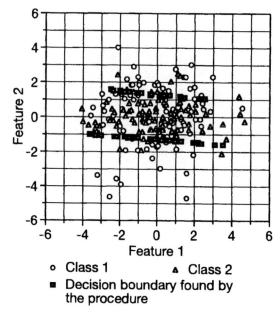

Figure 6-20. Samples and decision boundary for example 2.

Experiment 3 (Generated data)
(Two classes in three dimensions.) Generate 200 samples for each class with the following statistics.

$$\mu_1 = \begin{bmatrix} 0 \\ 0 \\ 0 \end{bmatrix} \quad \mu_2 = \begin{bmatrix} 0 \\ 0 \\ 0 \end{bmatrix} \quad \Sigma_1 = \begin{bmatrix} 3 & 0 & 0 \\ 0 & 3 & 0 \\ 0 & 0 & 1 \end{bmatrix} \quad \Sigma_2 = \begin{bmatrix} 1 & 0 & 0 \\ 0 & 1 & 0 \\ 0 & 0 & 1 \end{bmatrix}$$

$$P(\omega_1) = P(\omega_1) = 0.5$$

In this case there is no difference in the class means, and there is a covariance difference in only two of the three dimensions. Thus

304

the decision boundary should be a right circular cylindrical surface of infinite height, and just two features should be needed to achieve the same classification accuracy as obtained when using all three in the original space.

Upon applying the EDBFE algorithm, the following eigenvalues and eigenfunctions are obtained

$$\lambda_1 = 0.57581 \qquad \lambda_2 = 0.42032 \qquad \lambda_3 = 0.00387$$

$$\phi_1 = \begin{bmatrix} 0.86 \\ -0.050 \\ 0.001 \end{bmatrix} \qquad \phi_2 = \begin{bmatrix} 0.49 \\ 0.84 \\ 0.21 \end{bmatrix} \qquad \phi_3 = \begin{bmatrix} -0.21 \\ -0.18 \\ 0.98 \end{bmatrix}$$

$$\text{Rank}(\Sigma_{EDBFM}) \approx 2$$

The Table 6-2 shows the accuracy obtained with the derived features. As predicted, the best set of features to use is two-dimensional.

Number of Features	Decision BoundaryFeature Extraction
1	65.0%
2	70.0%
3	70.0%

Table 6-2. Percentage of accuracy for the features derived in example 3.

Example 4 (Real Data).

For this example, data from the Field Spectrometer System (FSS) was used. This system produces data in 60 bands from 0.4 to 2.4 µm. It was mounted in a helicopter and flown at 60 m altitude providing an IFOV of 25 m. There were two classes of data available, Winter Wheat (691 samples) and Unknown Crops (619 samples). Figure 6-21 shows the mean spectral response determined

305

from the samples for the two classes. Notice that there is a relatively large difference in means over a significant portion of the spectral range.

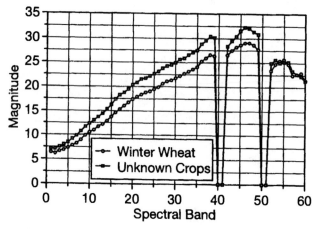

Figure 6-21. Mean spectral responses for the classes of example 4.

Figure 6-22 compares the performance for five different methods of feature selection. Uniform Feature Design is a simple combination of adjacent bands to achieve the desired number of features. For example, to form 30 bands from the 60 available, combine each pair of adjacent bands. The Foley and Sammon method[74] is another method that has appeared in the literature. It is known to perform well when there is a significant difference in mean values, but not so when there is not.

---

[74] D. H. Foley and J. W. Sammon, An optimal set of discriminant vectors, *IEEE Transactions on Computer*, Vol. C-24, No. 3, pp.281-289, March 1975.

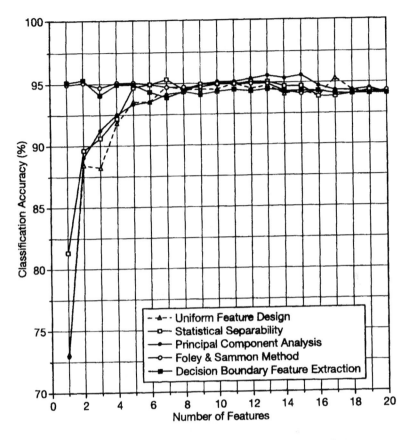

Figure 6-22. A performance comparison for example 4.

Table 6-3 shows the eigenvalues of the decision boundary feature matrix for various numbers of features, and compares them with the proportion of the total and the classification accuracy. Notice that the size of the eigenvalue at each dimensionality predicts very well the proportion of the total accuracy that would be obtained.

	Eigenvalues	Pro. Ev.(%)	Acc. Ev.(%)	Class. Acccracy(%)
1	0.994	49.6	49.6	93.4
2	0.547	27.3	77.0	94.3
3	0.167	8.3	85.3	94.4
4	0.133	6.6	91.9	95.0
5	0.066	3.3	95.2	95.1
6	0.041	2.1	97.3	94.9
7	0.020	1.0	98.3	94.9
8	0.012	0.6	98.8	94.8
9	0.008	0.4	99.2	95.0
10	0.007	0.3	99.6	95.3
11	0.005	0.2	99.8	95.3
12	0.001	0.1	99.9	95.7
13	0.001	0.0	99.9	95.5
14	0.001	0.0	100.0	95.4
15	0.000	0.0	100.0	95.3
16	0.000	0.0	100.0	95.6
17	0.000	0.0	100.0	95.5
18	0.000	0.0	100.0	95.5
19	0.000	0.0	100.0	95.4
20	0.000	0.0	100.0	95.4

Table 6-3. Accuracy for various numbers of features for example 4.

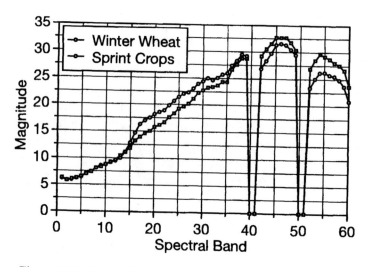

Figure 6-23. Mean difference between classes for example 5.

Example 5 (Real data)
As a final illustration, two different classes of FSS data were used. The classes were Winter Wheat (223 samples) and Spring Wheat (474 samples). Figure 6-23 shows that in this example there is much less difference in the mean vectors of the two classes. Figure 6-24 shows the performance comparison for the same five feature selection/extraction algorithms used in Example 5.

These examples illustrate that not only does the DBFE method work well in finding the best features, but it also predicts well the level of accuracy that will be achieved for a given dimensionality.

The Decision Boundary Feature Extraction process can involve quite lengthy calculations, but its chief limitation is that, because it

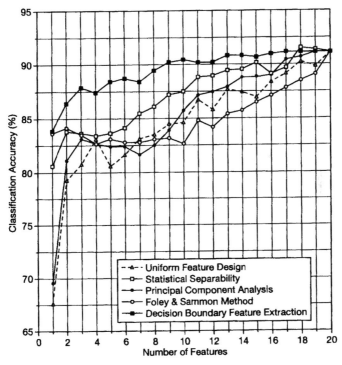

Figure 6-24. Performance comparison for example 5.

309

relies directly on training samples, it does not perform well when the training sets are small. Additional details on the Decision Boundary Feature Extraction algorithm, proofs of the theorems and lemmas, and a number of additional tests are available in the references.[75, 76] The method has also been generalized to apply to general nonparametric classification schemes and neural network implementations[77, 78]

6.7 Nonparametric Weighted Feature Extraction (NWFE)

A variety of conditions may affect the choice of a feature extraction algorithm for a given situation. Neither of the two optimal subspace determination methods for a classification task just described is ideal for all situations. DAFE has been the most widely used, because it works well in many situations. It is fast, but the limitations given at the end of Section 6.5–particularly that it produces only N-1 reliable features for an N class problem and that it is less effective when the class means are similar–limits its effectiveness in practical situations. DBFE, on the other had does not have these limitations, but it can involve lengthy calculations and more

---

75 Chulhee Lee and D. A. Landgrebe, Feature extraction based on decision boundaries, *IEEE Transactions on Pattern Analysis and Machine Intelligence*, Vol. 15, No. 4, pp. 388-400, April 1993.

76 Chulhee Lee and David A. Landgrebe, "Feature Extraction and Classification Algorithms for High Dimensional Data," PhD Thesis, December 1992 and Technical Report TR-EE 93-1, School of Electrical Engineering, Purdue University, West Lafayette, Indiana, January 1993.

77 Chulhee Lee and D. A. Landgrebe, Decision boundary feature extraction for non-parametric classification, *IEEE Transactions on System, Man, and Cybernetics*, Vol. 23, No. 2, March, 1993.

78 Chulhee Lee and D. A. Landgrebe, Decision boundary feature extraction for neural networks, *IEEE Transactions on Neural Networks*, Vol. 8, No. 1, pp. 75-83, January 1997.

significantly, because it is based directly on the training samples, it does not perform as well for small numbers of training samples.

Experience with DAFE and DBFE has led to the definition of a new feature extraction method called Nonparametric Weighted Feature Extraction (NWFE).[79] This method assumes the advantages of both DAFE and DBFE but mitigates their limitations. From DBFE, it was seen that focusing on samples near the eventual decision boundary location would be useful for improving DAFE. The main ideas of NWFE are putting different weights on every sample to compute the "local means" and defining new nonparametric between-class and within-class scatter matrices to get more features. In NWFE, the nonparametric between-class scatter matrix is defined as

$$S_b = \sum_{i=1}^{nc} \frac{P_i}{nc-1} \sum_{\substack{j=1 \\ j \neq i}}^{nc} \sum_{k=1}^{n_i} \lambda_k^{(i,j)} (x_k^{(i)} - M_j(x_k^{(i)}))(x_k^{(i)} - M_j(x_k^{(i)}))^T$$

where $x_k^{(i)}$ refers to the kth sample from class i. The scatter matrix weight $\lambda_k^{(i,j)}$ is defined as:

$$\lambda_k^{(i,j)} = \frac{dist(x_k^{(i)}, M_j(x_k^{(i)}))^{-1}}{\sum_{\ell=1}^{n_i} dist(x_\ell^{(i)}, M_j(x_\ell^{(i)}))^{-1}}$$

where dist(a, b) means the distance from a to b and $M_j(x_k^{(i)})$ is the local mean of $x_k^{(i)}$ in the class j. It is defined as:

79   Bor-chen Kuo, Improved statistics estimation and feature extraction for hyperspectral data classification, PhD thesis, Purdue University, December 2001.

$$M_j(x_k^{(i)}) = \sum_{\ell=1}^{n_i} w_\ell^{(i,j)} x_\ell^{(i)} x_\ell^{(j)}$$

where

$$w_l^{(i,j)} = \frac{dist(x_k^{(i)}, x_l^{(j)})^{-1}}{\sum_{l=1}^{n_j} dist(x_k^{(i)}, x_l^{(j)})^{-1}}$$

The nonparametric between-class scatter matrix is defined as

$$S_w = \sum_{i=1}^{nc} P_i \sum_{k=1}^{n_i} \lambda_k^{(i,i)} (x_k^{(i)} - M_i(x_k^{(i)}))(x_k^{(i)} - M_i(x_k^{(i)}))^T$$

The optimal features are determined by optimizing the criteria given by

$$J = tr(S_w^{-1} S_b)$$

To reduce the effect of the cross products of between-class distances and prevent the singularity, replace S_w by

$$S_w = 0.5 S_w + 0.5 diag(S_w)$$

Thus the NWFE algorithm is as follows:

Step 1. Compute the distances between each pair of sample points and form the distance matrix.

Step 2. Compute $w_l^{(i,j)}$ using the distance matrix

Step 3. Use $w_l^{(i,j)}$ to compute local means $M_j(x_k^{(i)})$

Step 4. Compute scatter matrix weight $\alpha_k^{(i,j)}$.

Step 5. Compute $\mathbf{S}_b$ and $\mathbf{S}_w$.

Step 6. Select the m eigenvectors of $\mathbf{S}_w^{-1}\mathbf{S}_b$, $\psi_1, \psi_2, \cdots, \psi_m$, which correspond to the m largest eigenvalues to form the transformation matrix $A_m = [\psi_1, \psi_2, \cdots, \psi_m]$

This algorithm appears to function better than the previous ones. Figure 6-25 shows a comparison of NWFE to DAFE on a 6-class problem with 40 training samples per class using 191 dimensional data. The data set involved was over a site dominated by agriculture/forestry cover types. Figure 6-26 shows the same type of comparison for an urban data set. In each case one can clearly see the limitation DAFE poses by creating only N-1 features in an N-class problem. These results also appear to show the value of weighting the training pixels near the eventual decision boundary more heavily than the pixels more distant from the boundary.

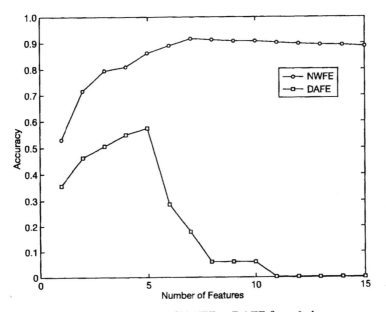

Figure 6-25. Comparison of NWFE to DAFE for a 6 class problem of 191 bands of data from an agriculture/forestry area using 40 training samples per class.

313

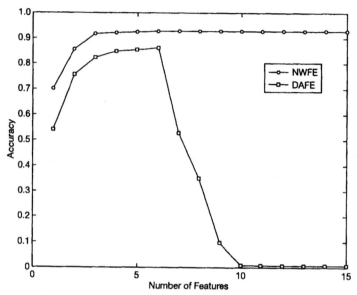

Figure 6-26. Comparison of NWFE to DAFE for a 7-
class problem of 191 bands of data from an
urban area using 50 training samples per class.

6.8 Projection Pursuit

The increased number of labeled samples required for supervised
classification as the dimensionality increases presents another
problem to current feature extraction algorithms where computation
is done at full dimensionality[80]. A method would be desirable that
computes in a lower-dimensional subspace instead of doing the
computation at full dimensionality,. Performing the computation
in a lower-dimensional subspace that is a result of a linear projection
from the original high-dimensional space would make the
assumption of normality better grounded in reality, potentially

---

[80] Lee, Chulhee and David A. Landgrebe, Feature extraction based on
decision boundaries," *IEEE Transactions on Pattern Analysis and
Machine Intelligence*, Vol. 15, No. 4, pp. 388-400, April 1993.

giving a better parameter estimation and better classification accuracy.

A preprocessing method of high-dimensional data based on such characteristics is based on a technique called Projection Pursuit. The preprocessing method is called Parametric Projection Pursuit.[81-83] Parametric Projection Pursuit reduces the dimensionality of the data maintaining as much information as possible by optimizing a Projection Index that is a measure of separability. The projection index that is used is the minimum Bhattacharyya distance among the classes, taking in consideration first- and second-order characteristics. The calculation is performed in the lower dimensional subspace where the data are to be projected. Such preprocessing is used before a feature extraction algorithm and classification process, as shown in Figure 6-27.

In Figure 6-27 the different feature spaces have been named with Greek letters in order to avoid confusion. Φ is the original high dimensional space. Γ is the subspace resulting from a class-conditional linear projection from Φ using a preprocessing algorithm, e.g. Parametric Projection Pursuit. Ψ is the result of a feature extraction method. Ψ could be projected directly from Φ

81   L. Jimenez and D. Landgrebe, Projection pursuit for high dimensional feature reduction: parallel and sequential approaches, presented at the *International Geoscience and Remote Sensing Symposium (IGARSS'95),* Florence Italy, July 10-14, 1995.

82   L. Jimenez and D. Landgrebe, Projection pursuit in high dimensional data reduction: initial conditions, feature selection and the assumption of normality, Presented at *IEEE International Conference on Systems, Man and Cybernetics (SMC 95),* Vancouver Canada, October 22-25, 1995.

83   L. Jimenez and D. Landgrebe, High dimensional feature reduction via projection pursuit, Technical Report TR-ECE 96-5, School of Electrical and Computer Engineering, April 1996, and Ph.D. thesis, Purdue University, May 1996.

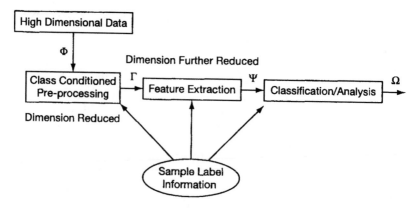

Figure 6-27. Classification of high dimensional data including reprocessing of high dimensional data.

or, if preprocessing is used, it is projected from Γ. Finally, Ω is a one-dimensional space that is a result of classification of data from Ψ space. The intention is to use Parametric Projection Pursuit in the role of Class Conditional Preprocessing, and a suitable class-conditional feature extraction method following this. Note that the three procedures—preprocessing, feature extraction and classification—all use labeled samples as a priori information.

<u>Illustrative Experiments.</u>

To show the relevance of high dimensional geometrical and statistical properties for high-dimensional data analysis purposes, two experiments were designed. In both experiments a comparison is provided between high dimensional feature extraction and the method that uses a Parametric Projection Pursuit based preprocessing to reduce the dimensionality before a feature extraction method is used.

The multispectral data used in these experiments are a segment of AVIRIS data taken of NW Indiana's Indian Pine test site. Of the original 220 spectral channels 200 were used, discarding the atmospheric absorption bands.

316

Experiment 1

In this experiment eight classes were defined. The total number of training samples was 1790 and the total number of test samples is 1630. The classification task for several classes in this and the next experiment are particularly difficult ones. The data were collected early in the growing season when the canopy of both corn and soybeans covered only about 5% of the area. There were three levels of tillage, no till in which there would be a great deal of residue on the soil surface from last year's crop, minimum till leaving a moderate amount of residue, and clean till for which there would be little or no residue. Added to this was the normal amount of spectral variability due to the varying soil types present in the fields. Thus the 95% background was highly variable, as compared to the relatively small difference in spectral response between corn and soybeans.

Classes	Training Samples	Test Samples
Corn-min	229	232
Corn-notill	232	222
Soybean-notill	221	217
Soybean-min	236	262
Grass/Trees	227	216
Grass/Pasture	223	103
Woods	215	240
Hay-windrowed	207	138
Total	1790	1630

Table 6-4. Classes and samples for the eight classes

Four types of dimension reduction algorithms were used. The first was Decision Boundary Feature Extraction (DB 200-22) to reduce the dimensionality from 200 bands to 22 features. The second was

Discriminant Analysis (DA 200-22) reducing the dimensionality again from 200 to 22. Both of these procedures performed a direct linear projection from Φ to Ψ. In the third and fourth methods Parametric Projection Pursuit was used to reduce the dimensionality from 200 to 22. These methods linearly project the data from Φ to Γ subspace. After that preprocessing method, a feature extraction algorithm was used to project the data once more from Γ to the Ψ subspace. Decision Boundary or Discriminant Analysis, (PPDB 22 and PPDA 22) offer the advantages of carrying out the computation with the same number of training samples in fewer dimensions.

Four types of classifiers were used. The first one was the quadratic ML classifier, the second was ML with 2% threshold. The third classifier was a spectral-spatial classifier named ECHO[84, 85] (see Chapter 8) and the fourth was ECHO with a 2% threshold. In the second and the fourth, a threshold was applied to the standard classifiers whereby in case of true normal distributions of the data, 2% of the least likely points would be thresholded. These 2% thresholds provide one indication of how well the data fit the normal model. All of these classifiers performed a projection from Ψ to the resulting space Ω.

The results are shown in Figure 6-28. The methods that use Parametric Projection Pursuit as a preprocessing method, with Γ as a stage between Φ and Ψ, performed better in terms of classification accuracy than methods directly using feature extraction at full dimensionality in Φ space (200 bands). That is

---

[84] R. L. Kettig and D. A. Landgrebe, Computer classification of remotely sensed multispectral image data by extraction and classification of homogeneous objects, *IEEE Transactions on Geoscience Electronics*, Volume GE-14, No. 1, pp. 19-26, January 1976.

[85] D. A. Landgrebe, The development of a spectral-spatial classifier for Earth observational data," *Pattern Recognition*, Vol. 12, No. 3, pp. 165-175, 1980.

because Parametric Projection Pursuit takes into account high-dimensional characteristics. Using feature extraction methods at full dimensionality can harm the data and impact the extraction of information, primarily because of the limited training data and the imprecise estimation of class statistics.

Observe in Figure 6-28 how greatly the performance of classifiers with 2% thresholds improves when using Parametric Projection

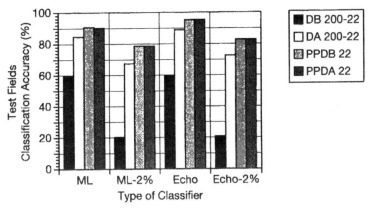

Figure 6-28. Test fields classification accuracy for four feature extraction methods and four classifiers.

Pursuit. This is because the computation is done at low dimensional space, Γ, so the assumption of normality is more valid. In the case of having fewer samples and classes Discriminant Analysis will be significantly affected by the high dimensional geometrical and statistical characteristics. The next experiment will show this difficulty.

Experiment 2

In this experiment four classes were defined: corn, corn-notill, soybean-min, soybean-notill. The total number of training samples is 179 (less than the number of bands used), and the total number of test samples is 3501.

319

Classes	Training Samples	Test Samples
Corn-notill	52	620
Soybean-notill	44	737
Soybean-min	61	1910
Corn	22	234
Total	179	3501

Table 6-5. Classes and samples for the four classes.

Two types of dimensional reduction algorithms were used. The first was Discriminant Analysis (DA 200-3) that reduces the dimensionality from 200 to 3. It directly projected the data from Φ space to Ψ subspace. In the second method Parametric Projection Pursuit was used to reduce the dimensionality from 200 to 22. It projected the data from the Φ space to the Γ subspace. After preprocessing was completed, Discriminant Analysis (PPDA 200-3) was applied to linearly project the data from the Γ subspace to the Ψ subspace. As mentioned before, this method provides an advantage of the computation being done with the same number of training samples but at lower dimensionality. The best three features were used for classification purposes, and the same four types of classifiers were used here as in the first experiment.

The results are shown in Figure 6-29. Parametric Projection Pursuit followed by Discriminant Analysis at lower dimensionality performed substantially better than using Discriminant Analysis at full dimensionality. The application of a threshold to Discriminant Analysis at full dimensionality reduced its classification accuracy more severely than when a threshold was applied where Projection Pursuit was first applied, followed by Discriminant Analysis at lower dimensionality. This is due to Parametric Projection Pursuit preprocessing being better fitted to the assumption of normality.

320

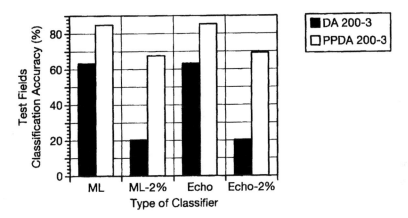

Figure 6-29. Test fields classification accuracy for two feature
extraction methods and four classifiers.

7 A Data Analysis Paradigm and Examples

In this chapter will be presented a paradigm for the analysis
of multispectral and hyperspectral data, bringing together a
number of the concepts and algorithms that were presented
in previous chapters into an overall data analysis process. This
will be followed by a series of analysis examples with various
data, illustrating some of these concepts and algorithms and
how they might perform in practice. The chapter is concluded
with a brief discussion of hierarchical classifiers.

7.1 A Paradigm for Multispectral and Hyperspectral Data Analysis

Recall the discussion at the close of Chapter 3, how one goes about
analyzing a multispectral image data set. Here we generalize the
ideas presented there to include some of the algorithms introduced
in more recent chapters that are suited for hyperspectral data as
well. The major question that the analyst must deal with is how to
choose and implement a suitable sequence of algorithms by which
to accomplish the desired analysis. Figure 7-1 outlines such a
sequence. This will be discussed in terms of the numbered boxes in
the Figure 7-1.

- Box 1. Multispectral data consists of data gathered in more than
  one spectral band. There is no accepted definition for where
  the boundary is between data termed multispectral and
  hyperspectral. However, in Chapter 5 it was established that
  the geometry of vector spaces changes continually as the
  dimensionality of the space increases, and indeed that it is
  materially different from the familiar three-dimensional
  geometry by the time dimensionality reaches 7 to 10. Further,
  it usually requires a dimensionality of the order of 10 or more

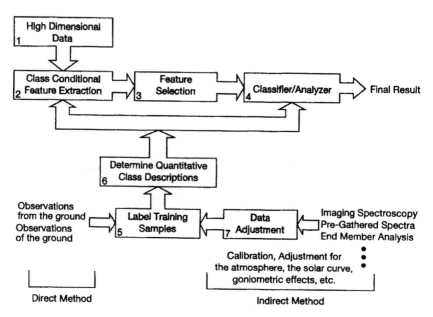

Figure 7-35. Tree structure for a Decision Tree designed by a hybrid method. The classes are Alfalfa (A), Corn (C), Oats (O), Red Clover (R), Soybeans (S), Wheat (W), Bare Soil (X), and Rye (Y).

to satisfactorily accomplish many practical analysis tasks. Thus it will be assumed that the data to be analyzed contains at least 10 and perhaps as many as several hundred spectral bands.

■ Box 2. Again assuming that the data were gathered in a larger number of bands than is necessary or desirable for the particular analysis at hand, an important early step is to form the feature set that is to be used in the analysis. The feature extraction should be done in a situation-specific way, using the description of the specific classes desired, derived from boxes 5 and 6. For example, this could be achieved by a classes-specific linear transformation such as DAFE, DBFE, and NWFE preceded by the projection pursuit technique described in chapter 6. A set of tentative features will be generateed that can be arranged in descending order of their usefulness in discriminating between the desired classes.

324

- Box 3. In step 2 there may be a decision to make as to how many of the generated features to utilize. The choice here and that in step 4, will depend to some extent on the individual classes and the precision with which they have been modeled.

- Box 4. There remains the question of what specific classification algorithm to use. Again, the choice of algorithm depends on the class model precision. Generally, the greater the precision, the more complex the algorithm that can be used.

- Box 5. The labeling of adequate sets of training samples is a key step, and perhaps the most important step of the entire process. In practical circumstances, there are a number of ways to obtain the necessary information to do this. This procedure will be discussed further below and illustrated in the examples to follow.

- Box 6. Having labeled a set of samples for each class, samples that are assumed to be truly representative of the desired classes as they exist in this data set, the task here is to use those samples to define as precise an N-dimensional model of the classes in the feature space as possible. Except in very simple cases where a single point in feature space is adequate, this will nearly always consist of modeling the entire distribution of each class. This may involve use of LOOC, an iterative scheme such as Statistics Enhancement, or it may simply consist of computing first and second order statistics. As discussed above, classes may require modeling in more than one mode, with the training samples divided between the various modes.

- Box 7. One option for labeling training pixels may be to adjust all or a part of the data for the various observational variables that were present, depending on the precise conditions of the scene and the sensor system at the time each pixel measurement was made. If one could do this adequately, this would make possible the use of some additional sources of reference data on which to base the labeling, as indicated by the diagram. The adjustment of the data for all of these variables is a very complex task and is problematic. It often cannot be done with as much precision as needed. Because of this, the overall scheme above is designed to not necessarily require calibrated data that has been so adjusted. Rather, one would only need to do so to the extent necessary to label an adequate set of training samples. If one has available information such as that indicated by one of the direct methods, the need for this added complexity can be avoided.

Clearly, the defining of classes and the labeling of a representative set of training samples such that an accurate class model can be calculated are the key activities for the analyst. There are a number of facets to this process:

- It is a fundamental fact of engineering that relative measurements are easier to make than absolute ones. In the remote sensing analysis context, this means that one must define an exhaustive list of classes so that there is a logical class to which to assign every pixel in the data set. Classifier algorithms that are of a nature of picking the best of all possibilities are such relative measurement schemes. They are likely to outperform those of an absolute nature, whose logic is simply to find all pixels that belong to a given class. If one is only interested in mapping a single class, it may be best to set up appropriate classes of "all other." In this way a required absolute decision can be converted into a relative decision, e.g., a classifier that can decide between the desired class and any of the others. Therefore, as has been previously emphasized, a fundamental requirement for establishing the list of classes to be used is that the list must be (a) exhaustive, (b) separable and (c) of informational value. Note that conditions (a) and (b) are conditions dependent on the data while the analyst determines (c).

- An important issue in the statistical approach to modeling the classes (box 6) is the question of whether a single mean vector and covariance matrix provide an adequate model for a class, regardless of how small the estimation error for these parameters can become. Using such a single set is equivalent to modeling the class as a single Gaussian random variable. Frequently classes have more complex structures than that, and more than one "subclass" may be established. The trade-off is that in such cases the limited number of training samples available must be divided among the several subclasses, thus increasing the likelihood of significant estimation error. In both of these two cases, clustering (unsupervised classification) may be of use in establishing an exhaustive list of classes with appropriate numbers of subclasses. A Proibability or Likelihood Map of a preliminary classification may also be used.

So far, a number of algorithms have been discussed that can be useful in this scheme. These algorithms and the box in Figure 7-1 where they apply are listed in Table 7-1.

Algorithm	Box
Discriminant Analysis Feature Extraction (DAFE)	2
Decision Boundary Feature Extraction (DBFE)	2
Nonparametric Weighted Feature Extraction (NWFE)	2
Projection Pursuit	2
Feature Selection	3
Maximum Likelihood Classification Algorithms	4
Clustering (Unsupervised Classification)	5
LOOC	6
Statistics Enhancement	6

Table 7-1. Algorithmsof previous chapters and their use in Figure 7-1

Next will be presented a series of examples showing how some of these algorithms are used in specific cases.

7.2 Example 1. A Moderate Dimension Example

Recall the Flightline C1 example described at the close of Chapter 3. Although the dimensionality of this data set does not really call for hyperspectral tools, it will be useful to see how additional algorithms perform in this moderate dimensional case before proceeding to several truly hyperspectral cases.

Using the same training and test data as in the example at the close of Chapter 3, the DBFE Feature Extraction algorithm was applied. One of the advantages of the DBFE algorithm is that it provides a direct indication of the number of the transformed features that should be used. Table 7-2 shows a portion of the results from applying DBFE. The size of the Eigenvalue shows in a subjective sense how significant that feature is in obtaining an effective separation of the classes defined. The entries in the "Percent" and "Cum. Percent" columns assist in that process by providing a normalization of the values. These numbers suggest that 4 or 5 of

328

the new features will be enough as will be seen, and this removes the need for the analyst to guess at how many to use.

Component	Eigenvalue	Percent	Cum. Percent
1	18.50	42.45	42.45
2	10.29	23.61	66.06
3	8.86	20.32	86.38
4	2.44	5.60	91.98
5	1.52	3.49	95.47
6	0.87	1.99	97.46
7	0.50	1.15	98.61
8	0.27	0.62	99.23
9	0.17	0.40	99.63
10	0.09	0.22	99.85
11	0.05	0.11	99.95
12	0.02	0.05	100

Table 7-2. Results of applying DBFE to the FLC1 training statistics.

The transformed data set was created from the DBFE transformation. The flightline was classified using the Quadratic Maximum Likelihood pixel classifier with this transformed data set using feature set sizes of the first 1 through the first 12. The training set and test set accuracies were determined for each. The results for test sample accuracy compared to that for the original, untransformed bands are shown in Figure 7-2. The results for the DBFE case are seen to be slightly better, but the small improvement may not be statistically significant. The DBFE algorithm has the advantage that its output provides information about how many features to use in making the classification. It also functions very satisfactorily for data sets of much larger dimensionality, and the advantage of such a transformation will be more apparent in the case of such problems.

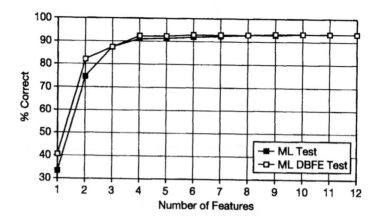

Figure 7-2. Accuracy vs. DBFE features

To improve the performance further, the flightline was classified
with DBFE features using the Quadratic ECHO spectral/spatial
classifier. This classifier combines the spectral information with
spatial information as will be described in the next chapter. The
results are compared with the Maximum Likelihood pixel
classification in Figure 7-3.

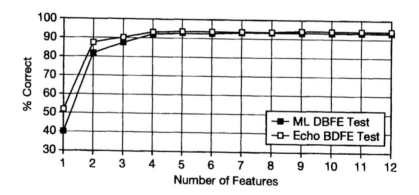

Figure 7-3. ECHO vs. pixel quadratic likelihood results using
the DBFE features.

As can be seen, in these tests the use of DBFE over the original spectral bands and ECHO over Maximum Likelihood pixel classification improves performance only marginally but consistently. Table 7-3 summarizes the results to this point. In a sense, this data set does not provide a severe enough test of the marginal improvement which these techniques may generally be expected to provide, because the classes are relatively separable. Only three or four features are required to achieve 90% accuracy or above.

	Quadratic		Quadratic DBFE		ECHO DBFE		Enhanced Quadratic		Enhanced ECHO	
	Train	Test	Train	Test	Train	Test	Train	Test	Train	Test
1	48	33	68	40	80	52				
2	89	75	92	82	99	87				
3	96	88	95	87	99	90				
4	97	91	98	93	99	94	98	93	99	95
5	98	92	99	93	100	94				
6	98	92	99	93	100	94				
7	98	93	99	93	100	94				
8	99	93	99	93	100	94				
9	99	93	99	93	100	94				
10	99	93	99	93	100	94				
11	99	93	99	93	100	94				
12	99	93	99	93	100	94	98	93	100	96

Table 7-3. Summary of all results.

A key characteristic of any such analysis is the generalization capability of the classifier. That is, how well does the classifier perform on samples other than its training samples, and how well does this generalization capability hold up over the entire data set. The use of test fields was designed to measure this characteristic to the extent that it can be done so quantitatively.

However, this generalization characteristic is very dependent on the analyst's selection of the training set. To assist in this process, the Statistics Enhancement algorithm was applied. As might be expected, this may sometimes result in a modest decline in the measured training set accuracy in favor of the broader test set, and this can often produce a better overall analysis.

The Statistics Enhancement algorithm was applied to the DBFE transformed data set and the data set was classified using the resulting enhanced statistics. Both the Quadratic Maximum Likelihood pixel classifier and the ECHO algorithm were used. The right four columns of Table 7-3 provide some results of this process. As can be seen that in this case, the quantitatively measurable accuracies remain at their previous levels or improve slightly. Figure 7-4 compares the data in image form with the enhanced statistics Quadric Maximum Likelihood and the enhanced statistics ECHO result.

One of the ways to quantitatively estimate the effect of the enhanced statistics is to compare before and after values of the Average Likelihood Probability, computed when a Probability Map is created during a classification. Table 7-4 lists results showing a substantial increase in the Average Likelihood Probability from its initial value for the Quadratic Maximum Likelihood classification to the final enhanced ECHO analysis as a result of the statistics enhancement process.

Number of	Average Probability (%)	
Features	ML	Enh.ECHO
4	19	36
12	21	35

Table 7-4. Average Likelihood for several parameter settings.

These results point to some of the possible processing steps in analysis of a multispectral data set and illustrate their likely impact in a typical situation. It is clear from such exercises that, no matter

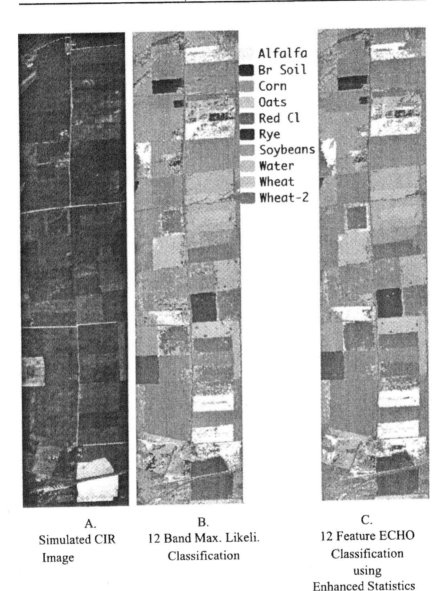

A.
Simulated CIR
Image

B.
12 Band Max. Likeli.
Classification

C.
12 Feature ECHO
Classification
using
Enhanced Statistics

Figure 7-4. The data set in image form and results from
analysis illustrating the effects of enhanced statistics. (In color on CD)

what algorithms are used to analyze multispectral data, the aspect of greatest importance is the accurate and thorough modeling of the classes of interest, relative to the other spectral responses that exist in the data set, and doing so in such a manner as to maximize the separability between them in feature space. This fact places great emphasis on the analyst and his/her skill and knowledge about the scene.

7.3 A Hyperspectral Example Exploring Limits and Limitations

Rather than an analysis for typical purposes, this example will demonstrate how far one might be able to go in discriminating between classes of a very subtle nature, if one can achieve a precise enough model of the desired classes. It will also illustrate what some of the primary limiting factors are to classification.

This example uses hyperspectral data containing 220 spectral bands from 0.4 to 2.4 μm. The data were collected by the AVIRIS airborne sensor system. The imaged area was of a small area (145 lines by 145 samples) gathered over the Indian Pines test site in Northwestern Indiana, a mixed agricultural/forested area, early in the growing season. Figure 7-5 shows a simulated color IR presentation of the data in image space.

The data set presents a very challenging classification problem. The primary crops of the area, corn and soybeans, were very early in their growth cycle with only about a 5% canopy cover. Further, different tillage practices were used in the fields, so different amounts of residue from the previous year's crop are on the surface. The differences in tillage practices are designated as follows:

- No till, a practice becoming steadily more common in which the new year planting is done without plowing under the residue of the previous year's crop. This

334

Figure 7-5 . Image Space display of the AVIRIS data using
bands 50, 27, 17 for red, green, and blue, respectively.
(In color on CD)

residue remains on the surface, protecting the soil from
erosion.

- Minimum till, in which only a minimum amount of field
preparation is done before the new crop is planted.
- Clean till, whereby the field is thoroughly plowed and
prepared prior to planting, so that only bare soil shows
on the surface.

As a result fields may be found with varying amounts of debris
from the previous year's crop showing on the surface in addition to

335

the natural spectral variation of soils in the fields. Figures 7-6 and 7-7 show examples of low percentage of canopy cover with the bare soil and debris debris evident in the background. Discriminating among the major crops under these circumstances thus presents a very challenging analysis problem.

Figure 7-6 . Close-up view of an area of a Corn-no til area, showing the low percentage of canopy ground cover present at the time of data collection. (In color on CD)

Figure 7-7. Close-up view of an area of a Soybeans-min. til field, showing the low percentage of canopy ground cover present at the time of data collection. (In color on CD)

To achieve discriminate successfully under these circumstances, a precise estimation of the class statistics is required in a rather large volume feature space, and therefore a large training set is required to reduce the estimation error as far as possible. Initially all of the labelable samples of the data set will be used for training. This does not allow for testing by an independent set, but the point here is to demonstrate that a set of classes with such subtle differences can be separated successfully in high enough dimensional space. Figure 7-8 shows the labeled areas that are used for estimating the class statistics.

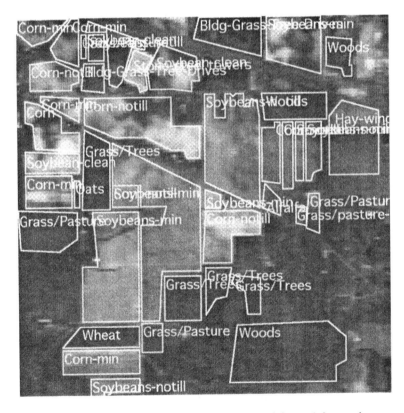

Figure 7-8. Labeled areas that will be used for training and evaluation. (In color on CD)

Table 7-5 lists the classes and subclasses used and the number of samples for each. If the task were to simply map corn and soybeans as the desired final classes and tillage practices were not of specific interest, then for reporting purposes to a user, the results for the three corn classes and for the three soybean classes could each be combined. This in effect models corn and soybeans each by three quadratic modes, and is a very effective way to handle non-Gaussian classes. Such an approach is referred to as a quadratic mixture classifier. It can handle class distributions of nearly any complexity. However, for the present purpose of exploring the potential of hyperspectral data of subtle classes, we will continue to treat all the listed classes individually.

Classes used:	Number of samples	LOOC Mixing
1: Alfalfa	54	1.88
2: Corn-notill	1434	1.38
3: Corn-min	834	1.53
4: Corn	234	1.78
5: Grass/Pasture	497	1.65
6: Grass/Trees	747	1.55
7: Grass/pasture-mowed	26	1.92
8: Hay-windrowed	489	1.64
9: Oats	20	1.98
10: Soybeans-notill	968	1.50
11: Soybeans-min	2468	1.16
12: Soybean-clean	614	1.63
13: Wheat	212	1.77
14: Woods	1294	1.31
15: Bldg-Grass-Tree-Drives	380	1.70
16: Stone-steel towers	95	1.76
Total	10366	
Average		1.63

Table 7-5. Training samples per class showing the LOOC mixing parameter value

It is noted that despite the use of most of the labeled areas for training, some of the classes do not have as many samples as there are spectral bands, so the LOOC algorithm is applied to obtain usable covariance matrices for all classes. Table 7-5 also lists the resulting values for the LOOC mixing parameter α. Notice that they are all between 1.0 and 2.0, which indicates that a combination of the ordinary sample covariance matrix and the common covariance matrix estimates is used. Note that for those classes with fewer training samples, the mixing is more nearly weighted toward the common covariance, as might be expected.

Figure 7-9 and Table 7-6 show the results of classifying the area using all 220 spectral bands with the LOOC-derived class statistics. The overall classification accuracy is 99.8% with only 22 of the 10,366 labeled samples incorrectly classified. However, some of the samples counted as errors are in fact not errors. For example, in Figure 7-8 it can be seen that there is a sod waterway in a soybean field in the lower left quadrant of the area. This waterway is labeled

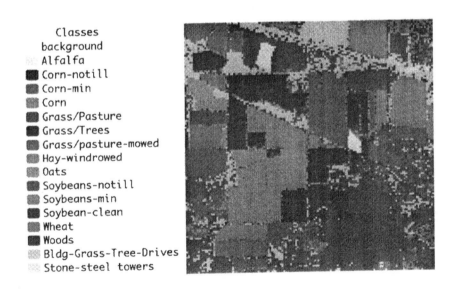

Classes
background
Alfalfa
Corn-notill
Corn-min
Corn
Grass/Pasture
Grass/Trees
Grass/pasture-mowed
Hay-windrowed
Oats
Soybeans-notill
Soybeans-min
Soybean-clean
Wheat
Woods
Bldg-Grass-Tree-Drives
Stone-steel towers

Figure 7-9 Thematic map resulting from the classification using LOOC statistics and all 220 bands. (In color on CD)

339

Grass

Class Name	Accur. %	No. pixels	Alfa.	Corn notill	Corn min	Corn	Grass Past	Grass Trees	pas. mow	Hay-Windr.	Oats	Soy-notill	Soy-min	Soy-clean	Wheat	Woods	Bldg Gr	Stone-tower
Alfalfa	100	54	54	0	0	0	0	0	0	0	0	0	0	0	0	0	0	0
Corn-notill	99.8	1434	0	1431	0	0	0	0	0	0	1	1	1	0	0	0	0	0
Corn-min	100	834	0	0	834	0	0	0	0	0	0	0	0	0	0	0	0	0
Corn	100	234	0	0	0	234	0	0	0	0	0	0	0	0	0	0	0	0
Grass/Past	100	497	0	0	0	0	497	0	0	0	0	0	0	0	0	0	0	0
Grass/Tree	100	747	0	0	0	0	0	747	0	0	0	0	0	0	0	0	0	0
Grass/Past-mowed	100	26	0	0	0	0	0	0	26	0	0	0	0	0	0	0	0	0
Hay-windrowed	100	489	0	0	0	0	0	0	0	489	0	0	0	0	0	0	0	0
Oats	100	20	0	0	0	0	0	0	0	0	20	0	0	0	0	0	0	0
Soybeans-notill	99.9	968	0	0	0	0	0	0	0	0	0	967	0	0	0	0	1	0
Soybeans-min	99.4	2468	0	2	0	0	10	1	0	0	0	2	2452	0	0	0	1	0
Soybean-clean	99.8	614	0	0	0	0	0	0	0	0	0	0	0	613	0	0	0	1
Wheat	100	212	0	0	0	0	0	0	0	0	0	0	0	0	212	0	0	0
Woods	99.9	1294	0	0	0	0	0	1	0	0	0	0	0	0	0	1293	0	0
Bldg-Grass-Tree-Dr	100	380	0	0	0	0	0	0	0	0	0	0	0	0	0	0	380	0
Stone-steel towers	100	95	0	0	0	0	0	0	0	0	0	0	0	0	0	0	0	95
TOTAL		10366	54	1433	834	234	507	749	26	489	21	970	2453	613	212	1293	382	96
PERCENT			100	99.9	100	100	98	99.7	100	100	95.2	99.7	100	100	100	100	99.5	99

OVERALL CLASS PERFORMANCE (10344 / 10366) = 99.8%

Kappa Statistic (X100) = 99.8%.   Kappa Variance = 0.000000.

Table 7-6. The accuracy matrix for the 220 band classification result shown in Figure 7-9.

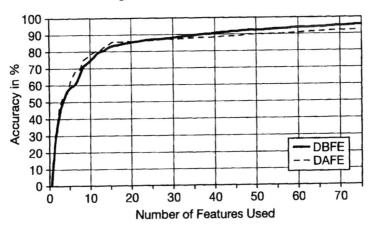

**DBFE and DAFE Feature Classification Accuracy
Using All Labeled Samples for Training**

Figure 7-10. Graph showing the training sample accuracy for various numbers of featured derived by DBFE and DAFE. Both curves continue to rise out to 220 features with no maximum, however, the DBFE rises a little faster than DAFE and finishes at 99.8% as compared to 99.6% for DAFE features.

as part of a soybean class, although the waterway pixels are classified as grass/pasture, and probably correctly so.

This classification rather clearly establishes that the defined classes do occupy separable volumes in feature space, and can be discriminantd successfully. However, the training used in this case was rather unusual in that one would not ordinarily expect to be able to use that large a portion of the area of the region to be classified as training. We will explore this point further shortly. Before doing so, we consider how the size of the feature space affects the separability in this case. Figure 7-10 shows the classification accuracy as a function of feature space dimensionality where the subspaces were derived by either DAFE or DBFE. Notice that DAFE appears to have a very slight advantage at lower dimensionality, while DBFE performs slightly better above about 20 features. Because of the relatively large training sets, the curves continue to increase all the way to 220 features and do not show any Hughes effect.

Next we explore the performance when the training sets are perhaps more realistic in size. To do so, the training areas are reduced to small areas of 4 x 4 pixels within each of the former training fields. This way, though samples from the same fields are used, both the

Figure 7-11. The small rectangles among the outlined labeled areas are the training areas used for the small training set test. It includes 688 samples in total. (In color on CD)

number of samples and the degree of representativeness of each training field is reduced. The areas thus chosen are shown as the small rectangles of Figure 7-11. In a few cases, where the labeled area had a dimension smaller than 4 pixels, the shape of the training area was adjusted so as to stay within the labeled area while maintaining the 16 pixel size.

The total training set size becomes 688 pixels, and the training set accuracy reaches 100% with 25 features, as seen in Figure 7-12. The training set accuracy does not decrease from there to 220 features. However, the test accuracy using all 10,366 labeled samples for test reaches a maximum accuracy of only 63.3% at 29 features and then slowly declines to 60.8% at 220 features. Clearly, the new small training set is not completely representative of the

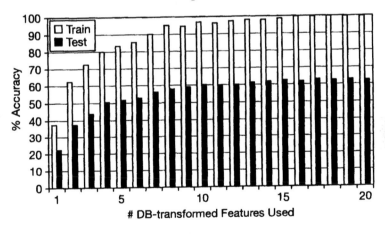

Figure 7-12 Classification results for a much reduced training set. The training set accuracy reaches 100% with 25 features and does not decrease from there to 220 features. The test accuracy using all 10,366 labeled samples reaches a maximum accuracy of only 63.3% at 29 features and then slowly declines to 60.8% at 220 features.

343

variations present in the classes, and given the beginning of the Hughes effect, there is some portion of the performance loss due to estimation error in the class statistics.

To explore this matter further, we will enlarge the training set by expanding the 4 x 4 pixel areas to 6 x 6 areas. This enlarges the small training set size from 688 pixels to an intermediate size of 1490 total pixels. Table 7-7 shows the training set size and LOOC mixing parameter values for each class. The mixing parameter average is shown to be larger than the value for the large training set given in Table 7-5, 1.85 vs. 1.63 for the 10366 sized training set, indicating that the covariance matrix is more heavily weighted toward the common class statistics, as would be expected for the smaller training set.

Classes used:	Number of samples	LOOC Mixing Parameter
1: Alfalfa	36	1.90
2: Corn-notill	212	1.73
3: Corn-min	180	1.75
4: Corn	40	1.88
5: Grass/Pasture	132	1.80
6: Grass/Trees	144	1.80
7: Grass/pasture-mowed	28	1.94
8: Hay-windrowed	36	1.91
9: Oats	18	2.00
10: Soybeans-notill	132	1.82
11: Soybeans-min	178	1.74
12: Soybean-clean	108	1.82
13: Wheat	36	1.91
14: Woods	108	1.81
15: Bldg-Grass-Tree-Drives	72	1.86
16: Stone-steel towers	30	1.90
Total	1490	
Average		1.85

Table 7-7. Nnumber of training samples per class and the LOOC mixing parameter values for the intermediate sized training set. The mixing parameter average is shown to be larger than the value for the large training set, indicating that the covariance matrix is more heavily weighted toward the common class statistics, as would be expected for the smaller training set.

Figure 7-13 shows the rise in performance as the feature set size is increased for the intermediate sized training set. In this case, the training set accuracy reaches 100% at 70 features and does not decline from that value. The test accuracy reaches a maximum of 72.4% at 80 features, then slowly declines to 70.4% at 220 features. The rise in test accuracy of about 10% suggests that this training set does indeed capture more of the variability that exists in the classes but the continuing Hughes effect indicates that there is still a problem with estimation error.

One might think that the Statistics Enhancement process would mitigate or eliminate the problem with representativeness, however, because of the small area chosen for this example so that

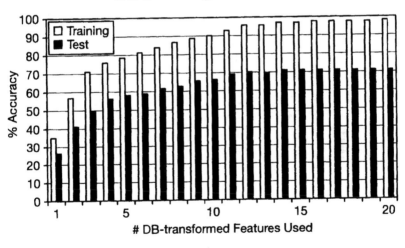

DBFE Transformed Classification Results
Medium Training Set - 1490 Samples

Figure 7-13. The results for an intermediate sized training set. The training set accuracy reaches 100% at 70 features and does not decline from that value after that. The test accuracy reaches a maximum of 72.4% at 80 features, then slowly declines to 70.4% at 220 features.

interpretations of performance can be clearly interpreted, some of the classes are so small that the statistics enhancement cannot be applied. To avoid singularity problems during the enhancement process, one must have at least one more training sample than the number of features. For several of the classes of this example, this is not possible.

Summarizing, we begin by noting that if there is any separability at all inherent in the spectral responses of the desired classes, it should be possible to achieve discrimination from data in a high dimensional feature space. The key parameters in achieving this discrimination are as follows:

- The precision of the estimates of the class statistics
- The representativeness of the training samples
- The optimal choice of the subspace in the feature space
- The optimal choice of a classifier

In this somewhat nontypical example, it was seen that:

- Given an adequately representative set of training samples that is large enough to reduce the estimation error to a negligible level, then a straightforward use of a quadratic classifier is effective, even for these very subtle classes. In this case, although performance could have been just as satisfactory for a subspace (reduced set of spectral features), feature extraction was not necessary because computation time was not a factor.
- For a small training set, both the representativeness and the estimation error became significant factors, with the former being more significant than the latter. An optimally chosen subspace is useful in this case.
- Enlarging the training set somewhat led to improved performance, apparently improving the representativeness more than the estimation error problem. Again, an optimally chosen subspace is useful in this case.

- From the results from the small and medium training set, the mixture values suggest that as the size of the training set is reduced, a Fisher linear discriminant (common covariance) classifier becomes more appropriate than one based entirely on class-conditional sample covariance estimates. There is no reason to assume that this will be true for every case, and the LOOC algorithm appears to be effective in determining the best mix of classifier algorithms to use.

7.4 A Hyperspectral Example of Geologic Interest

The characteristics of hyperspectral data can be exploited in a practical circumstance with the description of a hyperspectral data analysis task drawn from the field of geology.[86] Here, it is assumed that the analyst is interested in deriving a fairly detailed reconnaissance map with regard to a small number of specific minerals appearing on the surface. It is assumed that there are no observations from the ground available for the area. There is a generalized geological map for the area derived by conventional means available in the literature, but it will be used here only for checking the results, and not in the training process.

The area of interest is the Cuprite mining district in southwestern Nevada. The data, collected by a 1992 AVIRIS flight, consist of 220 spectral bands covering the range 0.4 to 2.5 μm. An image of one band of the data is shown in Figure 7-14. The site has several

---

[86] This example is taken from, J. P. Hoffbeck and D. A. Landgrebe, Classification of high dimensional multispectral data," PhD thesis and School of Electrical Engineering Technical Report TR-EE 95-14, May1995. See also J. P. Hoffbeck and D. A. Landgrebe, Classification of remote sensing images having high spectral resolution, *Remote Sensing of Environment*, Vol. 57, No. 3, pp 119-126, September 1996.

Figure 7-14. The 1.20 μm band from the Cuprite site.

exposed minerals of interest, including alunite, buddingtonite, dickite, illite, kaolinite, and quartz.

Part of the motivaton for high spectral resolution sensors such as AVIRIS is that the bands are narrow enough to coincide with molecular absorption features of individual molecules. As long as the region of exposed material on the surface is comparable to the size of a pixel, such absorption features can provide the means for labeling pixels.

One possibility might be to use knowledge of such absorption features to map the minerals directly. However, this would have the disadvantage of not allowing use of second-order statistics. Thus an alternate scheme will be used here. The knowledge of specific absorption featues will be used to label training samples. This way, a large enough set of training data can be found to enable using

348

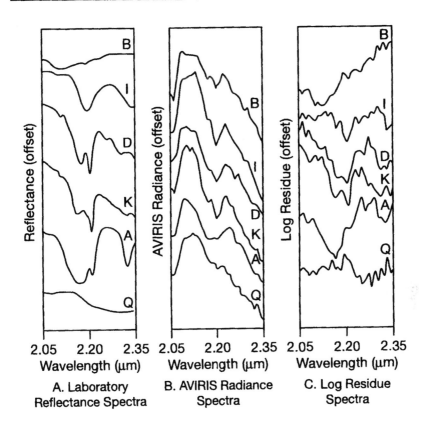

Figure 7-15. Reflectance, Radiance, and Log Residue Spectra (B - Buddingtonite, I - Illite, D - Dickite, K - Kaolinite, A - Alunite, Q - Quartz).

both first- and second-order statistics in the analysis. This approach also negates the need to use any kind of preprocessing scheme, such as atmospheric correction, and it thus avoids introducing any unnoticed or unanticipated deleterious effect that such preprocessing sometimes has, particularly on the second-order statistics.

The reflectance spectra of six minerals as measured in the laboratory are shown in Figure 7-15A. The wavelengths shown here represent only 31 of the 220 bands measured by the AVIRIS sensor. Each

349

mineral has absorption features where the reflectance curve reaches a local minimum. These absorption features will be used to label training samples for the classes of interest, however, there is a further problem to be overcome. The data for these graphs is for *reflectance*. AVIRIS data on the other hand, shows *radiance* spectra. Figure 7-15B shows the radiance spectra as measured in the Cuprite scene by the AVIRIS sensor of single pixels dominated by each of the six minerals. While the absorption features that distinguish the different minerals can easily be seen in the reflectance spectra measured in the laboratory, they can be quite difficult to see in the AVIRIS radiance spectra.

To overcome this problem, the log residue method[87] may be used to adjust the shape of the AVIRIS spectra to be more similar to the laboratory reflectance spectra. This adjustment suppresses multiplicative factors that are constant over the entire scene but vary with wavelength (e.g., the shape of the solar curve, atmospheric effects, etc.), and suppresses multiplicative factors that are constant with wavelength but vary over the scene (e.g., the effects of topology). The log residue calculation is based entirely on the data in the scene and does not require any external measurements of the atmospheric conditions. The log residue transformation is computed by the following formula:

$$y_{ik} = \frac{x_{ik} x_{..}}{x_{i.} x_{.k}}$$

where

y_{ik} – the log residue value of pixel and spectral band

---

[87] A. A. Green and M. D. Craig, Analysis of aircraft spectrometer data with logarithmic residuals. In *Proc. of the Airborne Imaging Spectrometer Workshop*, JPL Publ. No. 85-41, Jet Propulsion Laboratory, Pasadena, CA, pp. 111-119, 1985.

x_{ik} — the raw radiance data measured by the sensor

$x_{i.}$ — the geometric mean taken across the spectrum of pixel

$x_{.k}$ — the geometric mean of spectral band taken across all pixels

$x_{..}$ — the geometric mean taken across all pixels and all spectral bands

To avoid overflow problems, logarithms are used to compute the geometric means, and this gives rise to the name of the method. Although this method of adjustment is approximate, it is good enough for an analyst, with reasonable accuracy, to pick out AVIRIS pixels that have the desired absorption features and thus label them for use as training.

By this means, 1090 pixels of alunite, 71 pixels of buddingtonite, 162 pixels of dickite, 243 pixels of illite, 232 pixels of kaolinite, and 489 pixels of quartz were identified. Figure 7-16A shows the log residue of the average spectrum of the training pixels for each class.

Since dickite and kaolinite have very similar spectral features in the 2.05 – 2.35 μm range, it was not possible to accurately distinguish between these two minerals from the log residue spectra. Therefore training samples west of Highway 95 that resembled kaolinite and dickite were assumed to be dickite, and those to the east were assumed to be kaolinite. This assumption was based on a report that there was field verification of the existence of dickite on the west side of the highway but not on the east side, and kaolinite is known to exist on the east side.

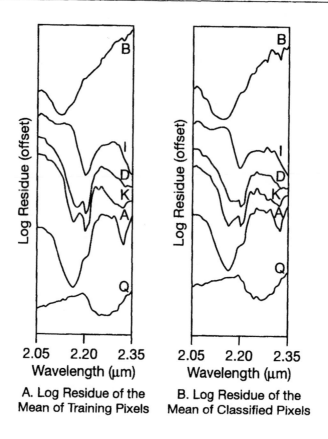

A. Log Residue of the
Mean of Training Pixels

B. Log Residue of the
Mean of Classified Pixels

Figure 7-16. Mean Log Residue Spectra of Training Pixels
and Classified Pixels (B-Buddingtonite, I - Illite, D -
Dickite, K - Kaolinite, A - Alunite, Q - Quartz).

Some pixels in the scene were mixtures of more than one mineral.
Since mixed pixels were not selected as training pixels, the final
classification represented the dominant mineral in each pixel. If
one were interested in mapping mixed pixels, additional classes
could be defined to represent various mixtures.

Discriminant analysis (DAFE) was performed with the six classes
with known absorption features and the 196 bands (0.40 − 1.35,

1.42 – 1.81 , and 1.95 – 2.47 μm) of the AVIRIS radiance data which were not in the water absorption regions of the spectrum. Equally likely classes were assumed. A Quadratic ML classifier was used to classify the resulting five discriminant features, and the likelihood map in Figure 7-17 was produced.

The dark areas of the likelihood map indicate samples that are unlikely to belong to any of the current training classes and suggest the location of additional classes. The spectra of several of these dark samples were examined to find more training samples of the six minerals. Additionally some areas were selected as training

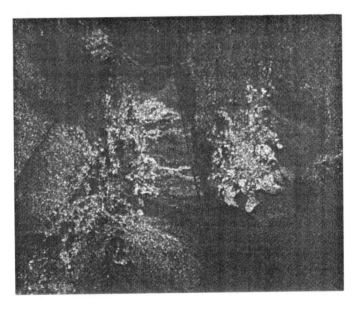

Figure 7-17. Likelihood Map from the preliminary classification. In the likelihood map, darker areas indicate low likelihood of membership in the class to which they have been assigned, while light areas indicate high likelihood.

353

samples for new classes of argillized, tuff, unaltered, and playa. Since classes like argillized, tuff, unaltered, and playa do not have known absorption features, additional information was required to determine their identity. In this case the names of the classes argillized, playa, and unaltered were determined from other information about the site in the literature. Had this information not been available, the training samples still could have been located and the classes mapped, but the names of the classes would not be known.

Next DAFE discriminant analysis was run using the 10 classes and 196 bands of the AVIRIS radiance data. To determine the appropriate number of features to use in the final classification, the classification accuracy was estimated by the leave-one-out method for the first discriminant feature, the first two features, the first three features, etc. It was found that using all nine discriminant features gave the highest leave-one-out accuracy.

The likelihood map in Figure 7-18 shows that the training samples represent the center of the scene reasonable well, but the areas near the edges may contain still additional classes. Figure 7-19 shows the classification of this data excluding two of the minor classes, and in a form that compares it to a conventionally derived map of the same area. It is seen that it compares very favorably with this map, but provides greater spatial detail, and is able to break out alunite and kaolinite as separate classes.

Classification accuracy is difficult to assess quantitatively without knowledge of the actual dominant mineralogy for each sample in the scene. Two indirect methods to estimate the accuracy of a classification are the resubstitution method, which is known to be optimistically biased, and the leave-one-out method, which provides

Figure 7-18. Likelihood Map for Final Classification

a lower bound on the accuracy. The resubstitution accuracy for the classification in Figure 7-19 was 96.7%, and the leave-one-out accuracy was 96.4%.

The log residue of the mean of the samples classified into each class (not including the training samples) is shown in Figure 7-16B. These curves compare reasonably well with the reflectance curves in Figure 7-16A, which lends support to the hypothesis that most of the samples are classified correctly.

Insensitivity to noise

The fact that this method of classifying high dimensional remote sensing data is relatively insensitive to noise is demonstrated by noting that the final classification changed very little regardless of whether or not the water absorption bands were included in the analysis. The reflected radiation in these bands (1.36 - 1.41 μm and 1.82 - 1.94 μm) is completely absorbed by the atmosphere, and so

Final Classification A   Map Using Previous Methods

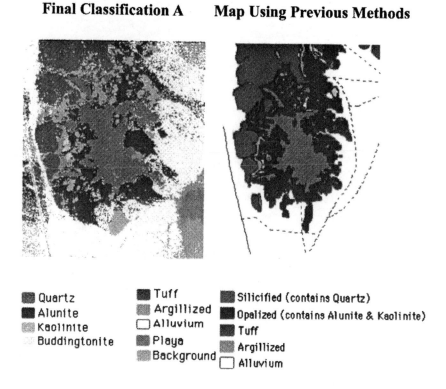

Quartz
Alunite
Kaolinite
Buddingtonite

Tuff
Argillized
Alluvium
Playa
Background

Silicified (contains Quartz)
Opalized (contains Alunite & Kaolinite)
Tuff
Argillized
Alluvium

Figure 7-19. Comparison of the final classification with a previous map appearing in the literature.[88] (In color on CD)

the data in these bands contained no information about the surface, only noise. When the water absorption bands were included in the analysis, the resubstitution accuracy was 97.0%, the leave-one-out accuracy was 96.7%, and the classification of only 4.75% of the samples changed.

88 Abrams, M.J., R.P. Ashley, L.C. Rowan, A.F.H. Goetz, and A.B. Kahle, 1977, "Mapping of Hydrothermal Alteration in the Cuprite Mining District, Nevada, Using Aircraft Scanner Images for the Spectral Region 0.46 to 2.36 μm," *Geology*, vol. 5, no. 12, pp. 713-718.

It is also instructive to examine the discriminant analysis features that resulted when the water absorption bands were included. Figure 7-20 shows the absolute value of the first four discriminant analysis features onto which the original data was projected. Each discriminant feature is a linear combination of all the original bands where the weights for each band are the elements of the discriminant features. The bands with the greatest weights are those near 2.15 μm, where the absorption features for some of the classes occur. This indicates that the discriminant analysis algorithm found these bands to be important in separating the classes. Furthermore, the bands in the water absorption regions were given relatively low weight. Figure 7-20 also shows the mean AVIRIS spectra over all

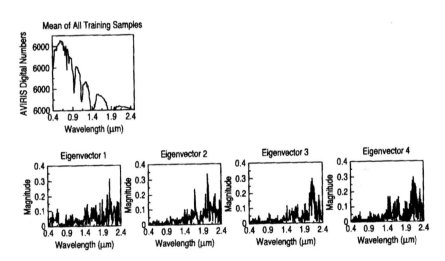

Figure 7-20. Absolute value of the discriminant analysis features, with the overall radiance vs. wavelength shown for comparison. The size of the discriminant analysis features at any wavelength provides some indication of the relative significance of that wavelength to the ability to discriminant between the classes. Note that the region including the absorption bands was found to be most significant, even though the irradiance level there was small, indicating a relatively poorer signal-to-noise ratio there.

357

the classes, and the water absorption regions (1.36 – 1.41 μm and 1.82 – 1.94 μm) are apparent.

An earlier analysis was performed on the 1987 AVIRIS data set taken over the same site. This data is known to have a much lower signal-to-noise ratio than the 1992 data set. The 1987 data set had been subjected to some radiometric adjustment at the time of its collection and consisted of 210 spectral bands evenly spaced 0.0098 μm apart covering the range 0.4000 to 2.4482 μm. Several training samples were identified by comparing the log residue spectra to the laboratory reflectance spectra in Figure 7-15. The number of training samples in each class was 236 (alunite), 65 (buddingtonite), 142 (kaolinite), 113 (quartz), 344 (tuff), 682 (unaltered), 192 (playa), and 80 (argillized). The original data were projected onto the first 8 discriminant analysis features and classified with a Quadratic ML classifier with a 0.1% threshold. Since the 1987 flight did not cover the area west of Highway 95, only the hydrothermal alteration zone to the east of the highway was analyzed. The resubstitution accuracy was 97.2% and the leave-one-out accuracy was 96.9%.

Conclusion

Currently spectral features that are diagnostic of some materials are usually detected by manual inspection or by a computer algorithm intended to identify the same features used in manual identification. These methods require that the remotely sensed radiance spectra be calibrated to reflectance in both magnitude and wavelength and have high signal-to-noise ratio. However, it is clear from signal theory principles that features that uniquely define the classes are already present in the uncalibrated radiance spectra, even though these features are not observable manually.

In this example, a method has been demonstrated for analyzing a data set of high spectral dimensionality. Such high-dimensional data not only make possible the use of narrow spectroscopic features where they are known to exist but are also able to make available the inherently higher information content of such data as predicted by signal theory principles. Training samples for materials with strong absorption features were located in the data using the log residue method to adjust the radiance spectra to subjectively resemble the reflectance spectra. Training samples for materials without strong absorption features or known reflectance spectra were located with other knowledge such as photo interpretation and ground observations. Features maximally effective in discriminating between the classes were then computed from the original radiance data by the discriminant analysis method, and these features were classified by a Quadratic ML classifier.

This method generated good results even with classes that lacked strong absorption features and with classes with similar spectral features. It effectively combined the human operator's knowledge of chemical spectroscopy with the power and robustness of a statistical classifier to perform the classification, greatly reducing the dependence of the analysis process on both reflectance and wavelength calibration and on high signal-to-noise ratio. This labeling ability of training samples by comparison to laboratory reflectance curves has the potential to greatly reduce the cost of labeling training samples for some classes.

Figure 7-21 shows the Statistics Images for four of the classes as a reference comparison.

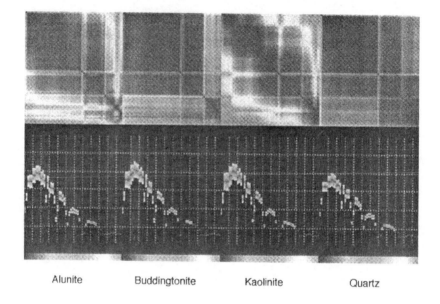

| Alunite | Buddingtonite | Kaolinite | Quartz |

Figure 7-21. Statistics Images for four of the classes of this example. (In color on CD)

7.5 Hyperspectral Analysis of Urban Data

This example is an analysis of an airborne hyperspectral data flightline over the Washington DC Mall. The data set contains 210 bands in the 0.4 to 2.4 μm region of the visible and infrared spectrum and has 1208 scan lines with 307 pixels in each scan line. It totals approximately 150 Megabytes. With data this complex, one might expect a rather complex analysis process, however, the procedure described earlier results in a simple and inexpensive means to do so. The steps used and the time needed on a personal computer for this analysis are listed in Table 7-8 and are briefly describe below.

Operation	CPU Time (sec.)	Analyst Time
Display Image	18	
Define Classes		< 20 min.
Feature Extraction	12	
Reformat	67	
Initial Classification	34	
Inspect and Add 2 Training Fields		≈ 5 min.
Final Classification	33	
Total	164 sec = 2.7 min.	≈ 25 min.

Table 7-8. Steps used and time required for the anaslysis

Step 1: Define classes. Begin by presenting to the analyst a view of the data set in image form so that training samples can be marked. A simulated color infrared photo form is useful for this purpose, done with bands 60, 27, and 17 used for the red, green, and blue colors, respectively. This is the color IR image shown in Figure 7-22. The desired classes are Roof, Road, Grass, Trees, Trail, and Water. A class called Shadow was added for completeness.

Notice in the image that the class Roof is spectrally quite diverse, and further that the class Road and some examples of Roof do not appear spectrally very different.

Figure 7-22. A simulated color IR image of a data set collected over the Washington DC Mall. Bands 60, 27, and 17 of 210 bands were used for this image space presentation. (In color on CD)

Step 2: Feature extraction. After designating an initial set of training areas, apply Discriminant Analysis Feature Extraction

361

(DAFE). The result is a linear combination of the original 210 bands to form 210 new bands that automatically occur in descending order of their value for producing an effective discrimination. From the output, it is seen that the first nine of these new features will be adequate for successfully discriminating among the classes.

Step 3. Reformatting: Use the new features defined above to create a 9-band data set consisting of the first nine of the new features, thus reducing the dimensionality of the data set from 210 to 9.

Step 4. Initial Classification: Having defined the classes and the features, next carry out an initial classification. The ECHO classification algorithm discussed in the next chapter is used. This algorithm is a maximum likelihood classifier that first segments the scene into spectrally homogeneous objects. It then classifies the objects.

Step 5. Finalize Training: An inspection of the initial classification result indicates that some improvement in the set of classes is needed. Select two additional training fields and add them to the training set.

Step 6. Final Classification: Classify the data again using the new training set. The result is shown in Figure 7-23.

Hyperspectral data provide the capability to discriminate among nearly any set of classes. Research has shown that, of all the variables to the data analysis process, the most important one is the size and quality of the classifier training set. There are a number of additional steps that could be taken to further polish the result, but the current result appears to be satisfactory for many practical circumstances.

Figure 7-23. Thematic map generated from the hyperspectral data set. (In color on CD)

7.6 Analyst Dependence and Other Analysis Factors

As has been seen, there are a number of factors that can influence the success of a given analysis. Key among them are the adequacy in size and the representativeness of the training set chosen by the analyst. Also important is the choice of the number of subclasses made by the analyst. These choices are dependent on the knowledge and skill of the analyst. Basically, following the dictum of the computer user's mantra of "garbage in, garbage out," one cannot expect good analysis results out of the analysis process regardless of what algorithms are used, if one does not provide an adequately precise and complete model of the classes desired in the result by doing a good job selecting the list of classes and subclasses and choosing training samples for them.

Thus the relationship of the analyst's role in the process to the algorithms chosen for use is a significant one. To investigate the relationship among these

363

variables, a test illustrating this will be described next. This test was conducted by having 12 two-person teams conduct an analysis of the same 12-band data set under prescribed conditions.[89] The data set involved is Flight Line C1, the same as was used earlier in this chapter and also in Chapter 3. Each team was instructed to design a classifier for the following user classes[90]: *corn, oats, soybeans, wheat, and forage/hay.* Though there was extensive ground truth available for the area, the analysts, who had received significant instruction on the fundamentals of multispectral analysis but were all analyzing multispectral data for the first time, were limited to using 1000 or fewer total design samples, regardless of the number of subclasses they chose to define. This limitation was intended to approximate the more realistic situation where ground truth is more limited. They were then to apply each of several classification algorithms of varying complexity and to measure the performance of each in terms of the accuracy achieved on the 70,635-pixel test set previously described in Chapter 3. These classifications were carried out using class statistics derived in the standard way, and then repeat using LOOC statistics, followed by statistics calculated from Statistically Enhanced LOOC statistics. The results from this are termed the Baseline Test.

Next, to test the effect of training sets of reduced size, the teams were instructed to recompute all three sets of statistics after having selected for training only the upper left single pixel, the 2 x 2 upper left pixels, the 4 x 4 upper left pixels and the 8 x 8 upper left pixels

---

[89] D. Landgrebe, On the relationship between class definition precision and slassification accuracy in hyperspectral analysis, International Geoscience and Remote Sensing Symposium, Honolulu, Hawaii, 24-28 July 2000.

[90] The teams were permitted to use as many subclasses for each of these user classes as they desired to model the density function complexity of any given class, so long as they stayed within the 1000 pixel limit for the total number of training samples.

of each training field they had selected for the Baseline Test and repeat the classifications in each case. This way the effect of training set size from very small to more normal sizes could be demonstrated. The results are shown in Table 7-10. The various discriminant functions and algorithms involved are repeated here from previous chapters as Table 7-9, so that one may compare the various computations involved.

Minimum Distance to Means $g_i(x) = (x - \mu_i)^T (x - \mu_i)$

Fisher's Linear Discriminant $g_i(x) = (x - \mu_i)^T \Sigma^{-1}(x\dagger - \mu_i)$

Quadratic (Gaussian) Classifier $g_i(x) = -(1/2)\ln|\Sigma_i| - (1/2)(x-\mu_i)^T\Sigma_I^{-1}(x-\mu_i)$

Correlation Classifier $$g_i(\mathbf{x}) = \left(\frac{\mathbf{x}^T\mu_i}{\sqrt{\mathbf{x}^T\mathbf{x}}\sqrt{\mu_i^T\mu_i}} \right)$$

Spectral Angle Mapper (SAM). $$g_i(\mathbf{x}) = \cos^{-1}\left(\frac{\mathbf{x}^T\mu_i}{\sqrt{\mathbf{x}^T\mathbf{x}}\sqrt{\mu_i^T\mu_i}} \right)$$

Matched Filter or

Constrained Energy Minimization $$g_i(\mathbf{x}) = \frac{\mathbf{x}^T C_b^{-1}\mu_i}{\mu_i^T C_b^{-1}\mu_i}$$

Table 7-9 (begun). Discriminant functions and algorithms involved in the test.

LOOC

$$C_i(\alpha_i) = \begin{cases} (1-\alpha_i)\text{diag}(\Sigma_i) + \alpha_i\Sigma_i & 0 \le \alpha_i \le 1 \\ (2-\alpha_i)\Sigma_i + (\alpha_i - 1)S & 1 \le \alpha_i \le 2 \\ (3-\alpha_i)S + (\alpha_i - 2)\text{diag}(S) & 2 \le \alpha_i \le 3 \end{cases}$$

$$\mu_i = \frac{1}{N_i}\sum_{j=1}^{N_i} x_{ij} \qquad \Sigma_i = \frac{1}{N_i - 1}\sum_{j=1}^{N_i}(x_{ij} - \mu_i)(x_{ij} - \mu_i)^T \qquad S = \frac{1}{L}\sum_{i=1}^{L}\Sigma_i$$

Statistics Enhancement

$$\alpha_i^+ = \frac{\displaystyle\sum_{k=1}^{N} P^c(i\mid x_k) + \sum_{k=1}^{N_j} P_j^c(i\mid z_{jk})}{N\left(1 + \dfrac{N_j}{\displaystyle\sum_{r\in S_j}\sum_{k=1}^{N} P^c(r\mid x_k)}\right)} \qquad \mu_i^+ = \frac{\displaystyle\sum_{k=1}^{N} P^c(i\mid x_k)x_k + \sum_{k=1}^{N_j} P_j^c(i\mid z_{jk})z_{jk}}{\displaystyle\sum_{k=1}^{N} P^c(i\mid x_k) + \sum_{k=1}^{N_j} P_j^c(i\mid z_{jk})}$$

$$\Sigma_i^+ = \frac{\displaystyle\sum_{k=1}^{N} P^c(i\mid x_k)(x_k - \mu_i^+)(x_k - \mu_i^+)^T + \sum_{k=1}^{N_j} P_j^c(i\mid z_{jk})(z_{jk} - \mu_i^+)(z_{jk} - \mu_i^+)^T}{\displaystyle\sum_{k=1}^{N} P^c(i\mid x_k) + \sum_{k=1}^{N_j} P_j^c(i\mid z_{jk})}$$

Table 7-9(concluded). Discriminant functions and algorithms involved in the test.

Classifier	Min Dist	Maximum Likelihood Classifiers			Special Purpose Classifiers	
		Fisher Lin. Disc	Quad. Pixel	Quad ECHO	Corr (SAM)	Matched Filter (CEM)
Baseline Standard (100-200 pixels/class)						
Ave	75.1%	86.9%	91.4%	92.8%	80.9%	69.8%
St. Dev.	9.1%	4.0%	1.9%	1.8%	6.0%	3.6%
LOOC Ave	75.1%	87.0%	92.0%	93.9%	80.8%	69.8%
St. Dev.	9.1%	4.0%	2.3%	2.3%	6.0%	3.6%
LOOC Enh. Ave	74.4%	86.9%	91.5%	94.6%	80.2%	64.1%
St. Dev.	7.6%	2.9%	2.5%	2.3%	5.9%	5.4%
1 Pixel/field						
Ave	75.4%	69.8%	76.4%	78.9%	79.7%	65.2%
St. Dev.	7.7%	13.2%	*	*	7.1%	6.2%
LOOC Ave	75.6%	81.9%	79.9%	84.0%	80.4%	65.5%
St. Dev.	8.0%	6.1%	6.1%	8.5%	7.3%	6.2%
LOOC Enh. Ave	69.3%	81.7%	83.6%	87.2%	76.1%	54.1%
St. Dev.	10.9%	4.5%	5.5%	6.1%	8.5%	9.6%
4 Pixel/field						
Ave	71.7%	83.6%	75.9%	76.0%	79.8%	64.5%
St. Dev.	18.0%	7.7%	6.6%	6.4%	5.9%	16.9%
LOOC Ave	71.6%	84.2%	84.0%	85.6%	79.9%	64.6%
St. Dev.	18.0%	7.8%	7.7%	7.3%	5.9%	17.0%
LOOC Enh. Ave	73.5%	83.1%	86.3%	88.0%	77.2%	61.5%
St. Dev.	8.8%	9.0%	8.8%	10.1%	9.8%	10.1%
16 Pixel/field						
Ave	75.7%	87.2%	88.6%	89.7%	81.0%	70.0%
St. Dev.	8.6%	3.5%	4.2%	4.7%	5.5%	3.6%
LOOC Ave	75.7%	87.2%	91.7%	93.6%	81.0%	70.8%
St. Dev.	8.6%	3.8%	2.2%	2.4%	5.5%	4.4%
LOOC Enh. Ave	73.3%	85.8%	91.0%	94.2%	78.1%	61.7%
St. Dev.	7.2%	4.2%	2.6%	2.1%	7.1%	4.6%
64 Pixel/field						
Ave	76.5%	88.3%	92.8%	94.6%	81.4%	70.3%
St. Dev.	7.7%	3.2%	1.1%	1.0%	5.1%	3.5%
LOOC Ave	76.5%	88.3%	93.1%	95.0%	81.3%	70.3%
St. Dev.	7.7%	3.2%	1.0%	0.9%	5.2%	3.5%
LOOC Enh. Ave	72.7%	84.9%	90.2%	93.2%	77.9%	59.4%
St. Dev.	6.4%	4.7%	4.5%	5.3%	7.3%	6.4%

Table 7-10. Average accuracy obtained and the standard deviation of that accuracy for the test of the 12 analysis teams.

367

The results in Table 7-10 show that for the baseline case and using standard statistics (the first row of results), the performance steadily improves through the first four columns. These four classifiers are a family of maximum likelihood classifiers using steadily increasingly detailed models of the individual class density functions. From Table 7-9, it is seen that the Minimum Distance classifier uses only the class mean values, the Fisher Linear Discriminant uses the class mean values and a common covariance matrix estimated using the training samples from all classes, and the Quadratic Maximum Likelihood (and ECHO) use individual class means and class covariance matrices. There are two additional algorithms tested, the Correlation or Spectral Angle Mapper, which like the Minimum Distance classifier, uses only the class mean values, and the Matched Filter or Constrained Energy Minimization classifier, which uses the class mean values and a covariance matrix estimated from the entire data set. In this test case, the Correlation or Spectral Angle Mapper provided performance in between that of the Minimum Distance Classifier and the Fisher Linear Discriminant, while the Matched Filter classifier, which is really intended for another purpose, provided the least attractive performance of all. Henceforth, attention will be focused on the four maximum likelihood classifiers.

Relative to sensitivity to the analyst, the figures for the standard deviation of performance for the 12 analysis teams, given in the line below that for the average performance for the 12 analysts, appear to be quite low, at least for the cases of the more powerful classifiers. This suggests that performance sensitivity is quite low when good analysis practices are followed. Use of LOOC and LOOC Enhanced statistics tended to provide a slight improvement in performance, although there are exceptions.

Regarding the variation with training sample set size, looking down the column for the Minimum Distance classifier, one sees very little change in performance, although the accuracy for all cases was not very high. This suggests that on the one hand, estimation of the class mean vectors is not very sensitive to training set size. On the other hand, the variation among analysts is significant, as indicated by the standard deviation of the accuracies over the 12 teams.

As one moves to the right, reviewing results when using more complex class models involving covariance matrices, it is apparent that training set size is significant. It is well known that the estimation error involving estimation of a covariance matrix is a significant function of the training set size. Thus it is seen that use of a more complex algorithm can provide improved results, but only if the estimation error of the parameters involved can be held to a small enough value.

7.7 Summary and Directions

The examples presented in this chapter were intended to sample the spectrum of possible analysis circumstances and applications, including examples from agriculture, geology, and an urban setting. It is clearly demonstrated here that one of the most significant factors in obtaining a satisfactory analysis result is the adequately precise specification of an exhaustive list of classes. Thus, on the one hand, an adequately large and representative set of training samples is of prime importance. On the other hand, defining and labeling training samples in a data set is one of the most onerous tasks in analysis.

One of the directions of current research[91] is to make this process simpler and more automated. Figure 7-24 shows one such line of work in progress. Box 8 has been added to the scheme of Figure 7-1, showing the use of samples initially classified to each of the classes as being added to the training set. These samples, referred

369

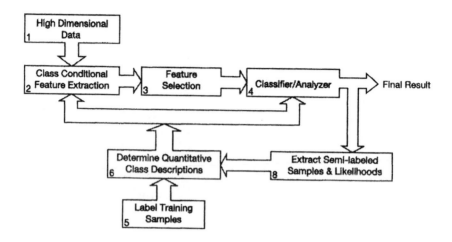

Figure 7-24. Organization of a tentatively augmented analysis system.

to as semilabeled samples since they are only tentatively members of a specific class at this point, are used to reduce the class parameter estimation error and the data are reclassified. Only samples that have a likelihood value above a threshold are added to the training

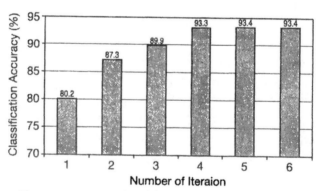

Figure 7-25. Classification accuracy for Flightline C1.

[91] Q. Jackson and D. Landgrebe, An adaptive classifier design for high-dimensional data analysis with a limited training data set, *IEEE Transactions on Geoscience and Remote Sensing.* Vol. 39, No. 12, pp. 2664-2679, December 2001.

set in this way. The data are then reclassified using statistics from the augmented training set. This process is repeated until convergence to a steady state value of performance is reached. Figure 7-25 shows the result of applying this scheme though six

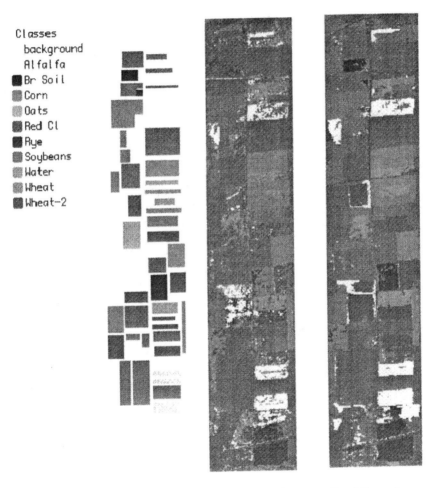

Classes
  background
  Alfalfa
  Br Soil
  Corn
  Oats
  Red Cl
  Rye
  Soybeans
  Water
  Wheat
  Wheat-2

Test fields   Initial Iteration   Final Iteration

Figure 7-26. Test fields and thematic map results for
FLC1. (In color on CD)

371

CONVENTIONAL CLASSIFIER

Sample

Water   Soil   Rock   Corn   Soybeans   Trees   Grass

Figure 7-27. Classification logic diagram for conventional classification.

iterations to the data of Flightline C1 of the first example. Figure 7-26 shows how the scheme affects the appearance of the result.

7.8 Hierarchical Decision Tree Classifiers

Decision Tree Classifier Concept and Advantages.

Another variable in the analysis process is the decision logic to be used. Figure 7-27 shows the conventional classifier logic as has been discussed up to the present. It is a single step process in which a sample is immediately assigned to a final class. However, sometimes, particularly in more complex analyses, it may be

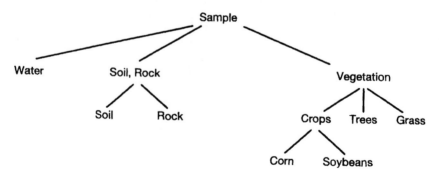

Figure 7-28. Possible hierarchical classifier logic for the task of Figure 7-27.

advantageous to use more complex classifier logic by introducing a hierarchical structure to the decision-making process, as illustrated in Figure 7-28.

There are possible advantages that may be gained by this more complex logic.

1. Providing a hierarchy to the classification process may make it easier to keep track of a large and complex class structure. This has historically been the method used when a large and complex body of information must be dealt with. Examples are the taxonomic system used by (1) classical biologists to sort out relationship among the plant kingdom, (2) the same for the animal kingdom, (3) the classification of the world's soil types, and (4) the Dewey decimal system used in library science.

2. It is possible to specialize the selection of features at each decision node to the particular discrimination that must be made there, so that some improvement in accuracy may be obtained.

It is also possible for the computation time to be reduced at the same time. This comes about because often smaller feature sets can be used at some nodes. A simple, hypothetical example may help clarify this, as follows: In a classifier that uses second order statistics, the computation time is proportional to the square of the number of features used. Figure 7-29 shows the case for a conventional, single level classifier. The variable c is the estimate of the computation time, which is simply the square of the number of features used.

However, if a decision tree approach is used as shown in Figure 7-30, and assuming the values of P at each node are the proportions of each class and the numbers of features indicated at each level

373

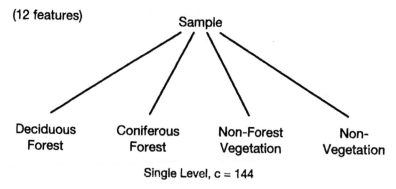

Single Level, c = 144

Figure 7-29. Hypothetical single-layer classification task using 12 features.

are used, then the computation time c would be as shown below the diagram. Thus, aside from any gain in accuracy that might be possible, a reduction by a factor of 144/53= 2.7 in the amount of computation required would be obtained in this case.

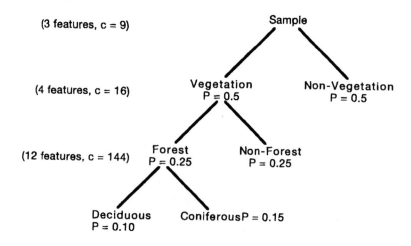

Multilevel, c = (1.0x9)+(0.5x16)+(0.25x144) = 53

Figure 7-30. The hypothetical classification of Figure 7-29 but using decision tree logic.

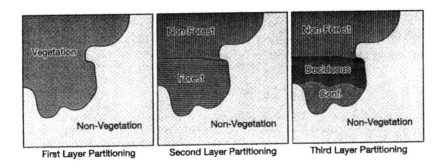

| First Layer Partitioning | Second Layer Partitioning | Third Layer Partitioning |

Figure 7-31. Effects in image space of layering classification.

Figure 7-31 shows the effect in image space of passing through the various levels of the decision tree for the illustration of Figure 7-30. It is seen that the process in this domain is one of sequentially dividing the geographical region into smaller and smaller, but increasingly detailed homogeneous areas. Figure 7-32 shows the sequential effect of the layered decisions in the feature space. Again, the process is one of steadily dividing the domain into smaller and smaller regions.

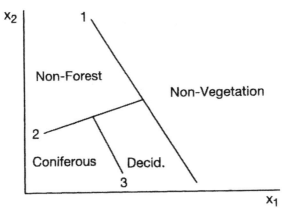

Figure 7-32. The effect in feature space of layering classification.

375

There is a price to be paid for the advantages of increased accuracy with decreased computation time, because the classifier design and training phases of the analysis process now become more complex.

Decision Tree Classifier Design

We will limit our discussion here to that of binary decision trees in which the decision between only two classes must be made at each node. This limitation has the advantage of being easier to optimize, and it has been shown that it can provide the same level of accuracy as a tree containing multiple classes at one or more nodes. The design question basically comes down to determining into what two subclasses the data are to be divided at each node and using what features. The question is then how to achieve this goal.

One could argue that the design procedure might be simply one of finding the two dominant modes of the data at each node using a suitable subset of the entire data, perhaps by a means such as clustering, and then picking the optimum feature set, i.e., the one for which these two modes have the greatest separability. We might refer to this approach as a *Top Down Approach*, since it begins at the top of the decision tree and proceeds toward the final list of classes at the bottom. However, as one proceeds through the various layers with this process, there is no guarantee that one will finally reach classes of interest or informational value.

Alternately, one could argue that the tree design process should begin at the bottom of the Decision Tree, with the training data that defines the classes of final interest. One can then successively group pairs of these classes together, proceeding upward toward the root node in defining the tree. However, by what rule does one pick the class pairs to combine, to guarantee maximum overall separability?

We have previously seen that the requirements for a suitably defined set of classes are that the set must be exhaustive, in that there must be a logical class with which to associate every pixel in the data set, the classes must be adequately separable in terms of the feature set available, and the classes must be of informational value (i.e. useful to the user). Furthermore these conditions must be simultaneously satisfied by the class list.

Thus, from the tree design standpoint, one can see how to approach each of these three requirements. The top down approach insures that the final classes reached will be separable, but not that they will be of informational value. The bottom up approach ensures that the final classes are of informational value, but not that they or those at the intermediate steps are separable. The challenge is to find a way to satisfy these requirements simultaneously.

A hybrid approach appears to be a useful approach. One may think of the concept as using the bottom up information to help steer the top down design. Figure 7-33 gives some results that show how such a process performs as a function of the number of features used at each node.[92] In this case the classification was made using Flight line C1 data and eight surface cover species as informational classes.

Figure 7-34 shows how the hybrid approach performs compared to a single layered classifier. Note that in both cases the highest accuracy was obtained using only a single feature at each node.[93] The resulting tree structure in this case is in Figure 7-35.

[92] Kim, B., and D.A. Landgrebe, "Hierarchical Classification in High Dimensional, Numerous Class Cases," (Ph.D. Dissertation), Technical Report TR-EE 90-47, Purdue University, Jun-90, 109 pp.

[93] Kim, B. and D.A. Landgrebe, "Hierarchical Classifier Design in High Dimensional, Numerous Class Cases," IEEE Transactions on Geoscience and Remote Sensing, Vol. 29, No. 4, July 1991, pp 518-528.

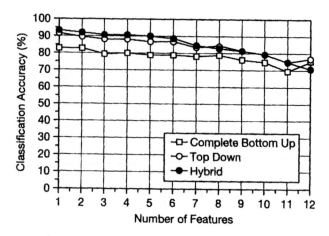

Classification Accuracy vs. Number of Features (8 Class)

Figure 7-33. Performance vs. dimensionality for several tree design approaches.

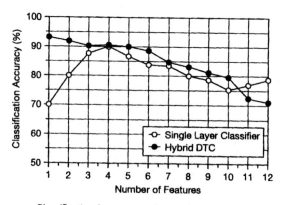

Classification Accuracy vs. Number of Features (8 Class)

Figure 7-34. Decision tree vs. single layered classifier performance.

378

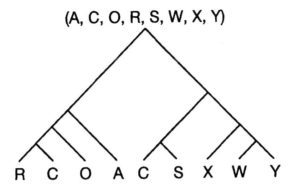

Figure 7-35. Tree structure for a Decision Tree de-
signed by a hybrid method. The classes are Alfalfa
(A), Corn (C), Oats (O), Red Clover (R), Soybeans
(S), Wheat (W), Bare Soil (X), and Rye (Y).

While useful in cases where the number of classes contained in the data are modest as was the case above, these more complex methods will be even more important with more complex data capable of producing separability for larger numbers of classes. Following are example results for a 23-class case. The resulting tree structure in this case is shown in Figure 7-37.

It is appropriate to conclude at this point, that although these methods show much promise, there is more research to be done to obtain useful analysis schemes for dealing optimally with high-dimensional, complex data situations.

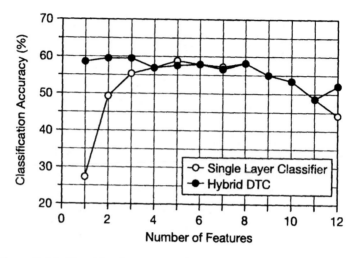

Figure 7-36. Classification Accuracy vs. Number of Features (23 Class)

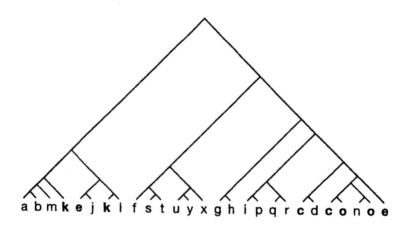

Figure 7-37. The tree structure for a 23 class problem. In this case, individual fields of the various crop species were taken as individual classes.

8 Use of Spatial Variations

In this chapter the use of spatial variations as an adjunct to spectral ones is presented. The chapter provides somewhat of a historical line of attack, beginning with a description of an early attempt to use a conventional, image processing approach by seeking a way to measure image texture. This is followed by a re-examination of the possibilities for the use of spatial variations, seeking one more suitable to multispectral image data. This leads to a description of the ECHO algorithm and a thorough test of its performance.

8.1 Introduction

A previously stated fundamental of remote sensing is that information is available at the pupil of the sensor based upon the spectral, spatial, and temporal variations of the electromagnetic fields emanating from the scene. Given that substantial information can be extracted from the spectral variations, the next question to be investigated is how to use the spatial variations in the analysis process.

8.2 Use of Texture Measures

Early thoughts on this matter quite naturally turned to studying how the human observer senses information from spatial variations, and this pointed very directly at scene texture. There have been many attempts to devise machine-implemented texture measures. We will study one in order to obtain a perception of some of the key aspects of texture measures and how they perform in general.

Chosen for study[94] is an early texture measure that is still occasionally referenced.[95]

The approach used to define a texture measure is by means of what are called Spatial Gray Tone Dependence Matrices, defined as follows:

  1. Define a sub-image or block of pixels of m x m pixels as the smallest object to be used. It is this sub-image in which a texture measure will be determined.

  2. The Spatial Gray Tone Dependence Matrix for each such sub-image is defined by the matrix $P = [p_{ij}]$ where p_{ij} is the number of occurrences of a pixel with gray tone i being a distance 1 from a pixel with gray tone j. If there are k possible gray tone values in the data, then P would be k x k dimensional

  3. Also for normalization purposes, define, R = number of possible neighboring cells.

A brief example will serve to illustrate this concept. Consider the following 5x5 block of data in which the number in each cell is the gray tone value of that pixel.

---

[94] R. M. Haralick, K. Shanmugam, and I. Dinstein, Textural features for image classification, *IEEE Transactions on Systems, Man, and Cybernetics*, Vol. SMC-3, No. 6, pp 610-621, November 1973.

[95] See, for example, A. Baraldi, and F. Parmiggiani, An investigation of the textural characteristics associated with gray level cooccurrence matrix statistical parameters, *IEEE Transactions on Geoscience and Remote Sensing*, Vol. 33, No. 2, pp 293-304, March 1995

Original Data	Re-Quantization	Re-Quantized Data
12 11 17 19 20	11–18 →1	1 1 1 2 2
15 17 19 20 22	19–22 →2	1 1 2 2 2
16 17 20 21 24	23–24 →3	1 1 2 2 3
14 20 22 23 24	25+up →4	1 2 2 3 3
21 23 24 26 25		2 3 3 4 4

Figure 8-1. Example preliminary calculation toward $\underline{P}$.

It is often useful to re-quantized the pixel gray values to a smaller number of gray values, as shown in Figure 8-1. This reduces the size of $\underline{P}$ to a more manageable size. Then one may form four directionally related Spatial Gray Tone Dependence Matrices, one each for the horizontal, vertical, left diagonal, and right diagonal, as follows:

$$P_H = \begin{pmatrix} 8 & 4 & 0 & 0 \\ 4 & 10 & 3 & 0 \\ 0 & 3 & 4 & 1 \\ 0 & 0 & 1 & 2 \end{pmatrix} \quad P_V = \begin{pmatrix} 10 & 3 & 0 & 0 \\ 3 & 10 & 4 & 0 \\ 0 & 4 & 2 & 2 \\ 0 & 0 & 2 & 0 \end{pmatrix}$$

$$R = 2N_y(N_x-1)$$

$$P_{LD} = \begin{pmatrix} 4 & 5 & 1 & 0 \\ 5 & 4 & 4 & 1 \\ 1 & 4 & 0 & 1 \\ 0 & 1 & 1 & 0 \end{pmatrix} \quad P_{RD} = \begin{pmatrix} 8 & 1 & 0 & 0 \\ 1 & 14 & 1 & 0 \\ 0 & 1 & 4 & 0 \\ 0 & 0 & 1 & 0 \end{pmatrix}$$

$$R = 2(N_y-1)(N_x-1)$$

Figure 8-2. Gray Tone Spatial Dependence Matrices for the data of Figure 8-1.

383

Examination of these matrices in light of their definition reveals how they manifest texture, as follows:

Course Texture $\rightarrow$ Neighboring Points Similar
$\underline{P}$ Large Near Diagonal

Fine Texture $\rightarrow$ Neighboring Points Different
$\underline{P}$ Relatively Uniform

Directional Texture $\rightarrow$ Coarser in one Direction
Degree of Spread About Diagonal Varies

Thus it is plausible at this point to hypothesize that these matrices have captured information about the texture present in the block from which they are derived. The next step in the process of developing a quantitative texture measure, then, is to devise a calculation with regard to the matrices that quantifies the degree of texture. Haralick, et al., proposed several for test, as follows:

- Angular Second Moment (ASM) – a measure of homogeneity

$$ASM = \sum_{i=1}^{N_g} \sum_{j=1}^{N_g} (\frac{P_{ij}}{R})^2$$

- Contrast (CONT) – a measure of the degree of local variation

$$CONT = \sum_{n=0}^{N_g-1} n^2 (\sum_{|i-j|=n} \frac{P_{ij}}{R})$$

384

- Correlation (CORR) – a measure of the gray tone linear dependence

$$CORR = \frac{\sum\limits_{i=1}^{N_g} \sum\limits_{j=1}^{N_g} ij(\dfrac{P_{ij}}{R}) - \mu_x\mu_y}{\sigma_x\sigma_y}$$

- Entropy (ENT) – a measure of the degree of variability

$$ENT = -\sum\limits_{i=1}^{N_g} \sum\limits_{j-1}^{N_g} (\dfrac{P_{ij}}{R}) \times \log(\dfrac{P_{ij}}{R})$$

where μ_x, μ_y, σ_x, and σ_y are the means and standard deviations of the marginal distributions associated with (i,j).

As an illustrative example, choose ASM for the example above:

$$ASM_H = \left(\frac{8}{40}\right)^2 + \left(\frac{4}{40}\right)^2 + \left(\frac{4}{40}\right)^2 + \left(\frac{10}{40}\right)^2 + \left(\frac{3}{40}\right)^2$$

$$+ \left(\frac{3}{40}\right)^2 + \left(\frac{4}{40}\right)^2 + \left(\frac{1}{40}\right)^2 + \left(\frac{1}{40}\right)^2 + \left(\frac{2}{40}\right)^2 = 0.1475$$

$$ASM_H = 0.1475$$
$$ASM_V = 0.1638$$
$$ASM_{LD} = 0.1172$$
$$ASM_{RD} = 0.2754$$
$$ASM_{AVE} = 0.1760$$

As a more specific example, if a 15 x 15 block size of LANDSAT MSS data from the 0.8 to 1.1 μm is used, the results are as follows:

	Agriculture	Water
ASM	0.0494	0.7641
Contrast	2.67	0.948
Correlation	0.514	0.545
Entropy	1.46	0.327

It can be observed from this that some of the measures appear to show good contrast for very different types of surface cover while others show less.

As a test of how these texture features perform, Haralick, et al. gave the following:

Sensor: Landsat 1 MSS
                 Monterey Bay, California region

Block Size: 64 x 64 pixels
                 0.6–0.7 µm band

Sample Size: 624 blocks (contiguous)
                 314 used for training set
                 310 used for test set

Classes: Seven Land Use -
                 Coastal Forest
                 Woodlands
                 Annual Grasslands
                 Urban
                 Small Irrigated Fields
                 Large Irrigated Fields
                 Water

True classes determined by photo-interpretation of color composite image

Features Used: Four Texture -
Angular Second Moment
Contrast
Correlation
Entropy

Eight Spectral - mean and variance
in each of the four spectral bands

The results of the test are shown in Table 8-1.

Class	Total Patterns	Coast. Forest	Wood land	Ann. Grass land	Urban Area	Lg. Irrig. Fields	Sm. Irrig. Fields	Water	% Correct
Coastal Forest	27	23	1	2	0	0	0	1	85%
Wood-land	28	0	17	10	0	1	0	0	61%
Annual Grass-land	115	1	3	109	1	1	0	0	95%
Urban	26	0	3	10	13	0	0	0	50%
Large Irrigated Fields	48	1	2	6	0	37	2	0	77%
Small Irrigated Fields	31	0	0	4	0	3	24	0	77%
Water	35	0	0	0	0	0	0	35	100%
TOTAL	310	25	26	141	14	42	26	36	

Table 8-1. Classification results from the texture features.

Overall, the results were as follows:

Overall Training Set Accuracy:	84%
Overall Test Set Accuracy:	83.5%
Overall Test Set Accuracy, Spectral Features Only:	74-77%

From these results, one might conclude that the texture features provided a useful assist to the spectral features in discriminating between these classes.

8.3 Further Evaluations of Texture Measures

As time went on, a number of other texture measures were put forth. A comparative study of a number of these subsequently appeared in the literature.[96] The texture features used in this comparison are as follows:

<u>Fourier Transform Texture Features</u>

---

[96] The following was taken from J. S. Weszka, C. R. Dyer, and A. Rosenfeld, A comparative study of texture measures for terrain classification, *IEEE Transactions on Systems, Man, and Cybernetics*, Vol. SMC-6, No. 4, pp 269-285, April 1976. An additional texture method evaluation appears in R. W. Conners and C. A. Harlow, A theoretical comparison of texture algorithms, *IEEE Transactions on Pattern Analysis and Machine Intelligence*, Vol. PAMI-2, No. 3, May 1980. The conclusions reached in this latter case are consistent with Weszka, et al. See also P. Gong, D. J. Marceau, and P. J. Howarth, A comparison of spatial feature extraction algorithms for land-use classification with SPOT HRV Data, *Remote Sensing of Environment*, Vol. 40, No. 2, May 1992. A current reference for such methods is K. M. S. Sharma and A. Sarkar, A modified contextural classification technique for remote sensing data, *Photogrammetric Engineering and Remote Sensing*, Vol. 64, No. 4, pp. 273-280, April 1998.

- Based upon Fourier Power Spectrum $|F|^2 = FF^*$ where

$$F(u,v) = \int_{-\infty}^{\infty} \int_{-\infty}^{\infty} f(x,y)e^{-j2\pi(ux+vy)}dxdy$$

- Ring Features based on

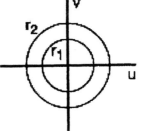

$$\Phi_{r_1 r_2} = \int_{r_1}^{r_2} \int_{0}^{2\pi} |F(r,\theta)|^2 d\theta dr$$

$$r^2 = u^2 + v^2 \quad q = \tan^{-1}\frac{u}{v}$$

Sensitive to coarseness

- Wedge Features based on

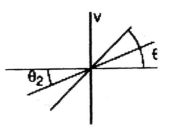

$$\Phi_{\theta_1\theta_2} = \int_{\theta_1}^{\theta_2} \int_{0}^{\infty} |F(r,\theta)|^2 dr d\theta$$

Second-Order Gray Level Statistics

These are simply the features used by Haralick, et al.

Gray Level Differences Statistics

Let $f_\delta(x,y) = |f(x,y) - f(x+\Delta x, y+\Delta y)|$
    where $\delta = (\Delta x, \Delta y)$

Assume that f_δ has m gray levels.

Let $p_\delta(i)$ be the probability density function of f_δ
where $0 \le i \le m$

The interpretation of f_δ is as follows:

> Texture Coarse $\to$ Neighboring values similar:
> f_δ usually small $\to$ p_δ concentrated near $i = 0$
>
> Texture Fine $\to$ Neighboring values different:
> f_δ often large $\to$ p_δ more widely spread
>
> Texture Directional $\to p_\delta$ varies with direction of δ

Then define the following features:

- Contrast $\quad CON = \sum i^2 \, p_\delta(i)$

- Angular Second Moment $\quad ASM = \sum p_\delta(i)^2$

- Entropy $\quad ENT = -\sum p_\delta(i) \, \log p_\delta(i)$

- Mean $\quad MEAN = \dfrac{1}{m} \sum i \, p_\delta(i)$

Gray-Level Run Length Statistics

Define a set of gray level ranges: $(0,a), (b,c), \dots (l, m)$. Let $p_\theta(i,j)$ be the number of runs j, in the θ direction, consisting of points whose gray levels lie in the ith range. The rationale in this case is the following:

- Texture Coarse $\to$ Neighboring points similar:
  Long runs frequent

- Texture Fine $\rightarrow$ Neighboring points different:
  Long runs less frequent

- Texture Directional $\rightarrow$ Run length dependence directional

Then define the following features:

- Long Run Emphasis: $\mathsf{LRE} = \dfrac{\Sigma j^2 p(i,j)}{\Sigma p(i,j)}$

- Gray Level Distribution: $\mathsf{GLD} = \dfrac{\sum\limits_{i}[\sum\limits_{j} p(i,j)]^2}{\sum p(i,j)}$

- Run Length Distribution: $\mathsf{RLD} = \dfrac{\sum\limits_{j}[\sum\limits_{i} p(i,j)]^2}{\sum p(i,j)}$

- Run Percentage: $\mathsf{RPC} = \dfrac{\sum p(i,j)}{\mathsf{N}^2}$ for an NXN pixel picture

The Weszka, et al. report comparisons of the texture measures based on the following data:

- Landsat 1 MSS data over Eastern Kentucky
- 180 samples each of 64 x 64 pixel windows
- 3 Geoterrain classes:
  Mississippi Limestone + Shale

Pennsylvania Sandstone
Lower Pennsylvania Shale
- Fisher linear discriminant classifier
- All data used for training

The results obtained are as follows.

	Fourier	2nd ord. CON points	2nd ord. CON ave.	Difference points		Difference averages	
				MEAN	CON	MEAN	CON
Max Correct	133	136	136	135	136	135	136
Min Correct	77	68	86	71	68	103	87
Ave Correct	102	108	113	108	108	114	113
Ave % Correct	57	60	63	60	60	63	63

Table 8-2. Number Correctly Classified using Individual Features.

	Fourier	2nd ord. CON points	2nd ord. CON averages	Difference points		Difference averages	
				MEAN	CON	MEAN	CON
Max Correct	158	166	166	167	166	165	166
Min Correct	151	160	163	164	160	163	158
Ave Correct	154	164	163	165	163	164	162
Ave % Correct	86	91	90	92	91	91	90

Table 8-3. Best Performing Pairs of Fourier, CON and MEAN Features

The following conclusions may be drawn from the results in Tables 8-2 and 8-3.
- Features based upon Second Order Statistics and Gray Level Statistics perform about equally,
- Fourier Features do not do as well,
- Results based upon small displacements appear better,
- The small size of the data set and the lack of generalization test leave the question of robustness (and thus practicality) open.

8.4 A Fresh Look at the Problem

The results obtained by the approach above were positive, but not overly so. It is seen that this approach does not differ much from that used in general image processing, in that it is directed at implementing an operation that a human finds useful in manual image analysis. This may or may not be appropriate, since the intention is to have a machine-implemented algorithm. For example, the approach above does not easily utilize multiple bands at a time, it certainly does not utilize interband information, and it is computationally quite intensive. It is perhaps useful to step back from the problem and take a fresh look at how spatial information of a scene is expressed in data in a manner that might be especially useful for a machine-implemented algorithm for multispectral data.[97]

Consider a vertical view of the Earth, as shown in Figure 8-3. What about the view carries usable information? It is seen that this appears to be an agricultural area, and one can easily find areas in the data (i.e., fields) that, while the class of ground cover is unknown, all pixels in the area seem very likely to belong to the same class.

8.5 The Sample Classifier Concept

This observation led to the concept of the Sample Classifier (as compared to a Pixel Classifier), where the term "Sample" implies a collection of points. The concept is more clearly expressed in Figure 8-4.

---

[97] What follows is an outline of a development effort conducted over a period of years and described in D. Landgrebe, The development of a spectral-spatial classifier for earth observational data, *Pattern Recognition,* Vol. 12, No. 3, pp. 165-175, 1980.

Figure 8-3. A vertical view of an Agricultural Area

In the case of the Pixel Classifier, one must decide to which of the two classes the single unknown, x, belongs. In the Sample Classifier, the unknowns, x, are all presumed to belong to the same class, and thus one must decide if the distribution of the unknowns has greater similarity to class 1 or 2. Thus, in the former case, one must compare a point to a set of density functions, while in the latter, one compares a density to a set of densities. Note also that this comparison is done in feature space, more suited to the use of machine implemented pattern recognition.

To illustrate why the density-to-density comparison is likely to provide more accurate results in the aggregate, consider two Gaussian classes in only one dimension. Let both classes have

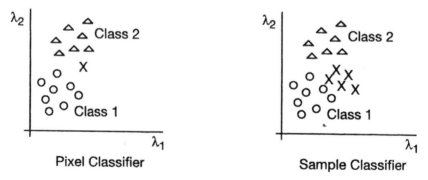

Figure 8-4. The sample classifier concept.

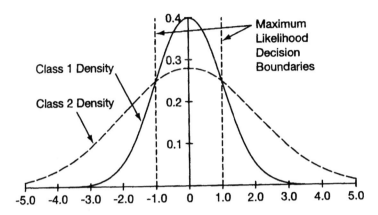

Figure 8-5. Two Univariate Normal Class Densities with Low Separability

identical mean values, but let $\sigma_1 = 1$ and $\sigma_2 = 2$. Then the conditional density functions would be as in Figure 8-5, and clearly, the error rate for a pixel classifier would be high. All pixels of class 1 for which $|x| > 1$ and all class 2 pixels for $|x| < 1$ would be classified in error.

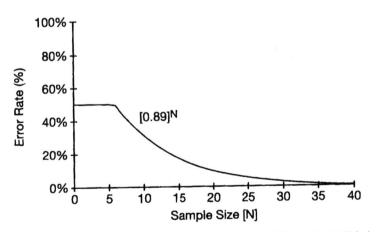

Figure 8-6. Error rate vs. sample size for the classes of Figure 8-5. This is a bound on the conditional probability of error for the maximum likelihood strategy as a function of sample size.

On the other hand, a sample classifier would be expected to do better, because the sample variance could be estimated and compared with the two classes. Clearly, if the sample size, N, were increased, an increasingly precise estimate of the variance could be obtained, and the error rate expected from a sample classification would decline to zero. The graph of Figure 8-6 shows one estimator of the expected error rate vs. sample size.

Notice that this means for classification is, indeed, multivariate in nature, and thus makes use of the inter-band information.

8.6 The Per Field Classifier[98, 99]

The above concept gave rise to the so-called "Per Field classifier," in which, with the usual training class statistics available, if one would manually designate the boundary of a given field, the pixels within the field could be classified based upon a minimum distance measure. The graphs of Figures 8-7, 8-8, and 8-9 show some comparative results obtained on multispectral data. Class statistics were derived by the usual means. The pixels within each field were then classified using the standard maximum likelihood pixel classifier and using the new method utilizing the Jeffreys-Matusita distance measure. In a further comparison a voting procedure was used as follows: if 60% or more of the pixels in a given field were classified to a given class using the pixel classifier, then it was

---

[98] T. Huang, Per-Field Classification of Remotely Sensed Agricultural Data, *Proceedings of the 1970 Allerton Conference on Circuit and System Theory.* LARS Information Note 060569, Purdue University, West Lafayette, Indiana, 1969.

[99] K-S. Fu, Pattern Recognition in Remote Sensing of the Earth's Resources, *IEEE Transactions on Geoscience Electronics,* Vol. GE-14, No. 1, pp 10-18, January 1976.

presumed that the whole field should be assigned to that class, and the error rate was computed accordingly.

These results suggest that the combination of spectral and spatial variations do indeed provide higher accuracy than spectral alone, and the concept warranted further investigation.

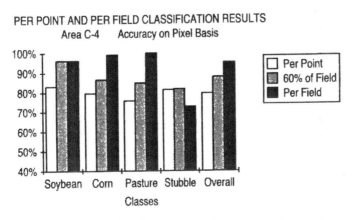

Figure 8-7. Comparison of Per Point and Per Field Classification Accuracy – Area C-4

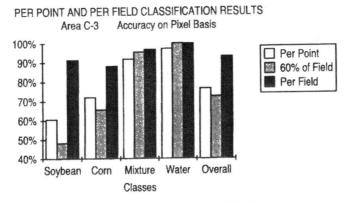

Figure 8-8. Comparison of Per Point and Per Field Classification Accuracy – Area C-3

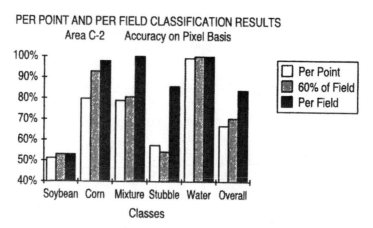

Figure 8-9. Comparison of Per Point and Per Field Classification Accuracy – Area C-2

8.7 Finding the Boundaries of Fields

But, of course, one would like to have a machine-implemented way to find the field boundaries, rather than doing so by manual means. Conventional image processing means, such as use of a Laplacian operator, could be used, and were tried.[100] However, these methods do not provide a means for capturing interband characteristics, and interband characteristics have been seen to be very significant.

A somewhat more powerful method might be to use multivariate clustering. To illustrate, suppose on the left of Figure 8-10 is the result of having clustered a small portion of multiband data. Then this cluster class map could be scanned using another algorithm to

---

[100] P. E. Anuta, Spatial registration of multispectral and multitemporal digital imagery using fast fourier transform techniques, *IEEE Transactions on Geoscience Electronics*, Vol. GE-8, No. 4, pp. 353-368, October, 1970.

```
4222222222222222222222222221222111
2222222222222222222222222222222111
222222222222222222222222222222221
22222222222222222222222222222222221        X - --- ------ - XXXXXXXX
22222422122222222222121111111111       X*-IX- ---------- -
33322244444424444444444444444444         -- X    -
11111112444444444444444444444444          X
11111112444444444444444444444444          X
11111112444444444444444444444444          X
11111114444444444444444444444444          II
11111112444444444444444444444444          X - -------------------
11111124343333333333333333333333          X - -------------------
11111113333333333333333333333333          I*- -
11111113333333333333333333333333          II
11111113333333333333333333333333          II
11111113333333333333333333333333          II
11111113333333333333333333333333          II
11111113333333333333333333333333          II
11111113333333333333333333333333          II
11111113333333333333333333333333          II
11111113333333333333333333333333          II
```

Figure 8-10. Finding field boundaries by clustering.

decide which pixels were vertical, horizontal, or mixed boundary pixels, with the results as shown on the right in the figure.

Figure 8-11 shows the result of applying this technique to a flightline of airborne multispectral scanner data.[101] The method clearly does locate boundaries in the data, and indeed multispectral boundaries. Further it is possible to arrange the algorithms such that various levels of thresholds for boundaries may be set. Figure 8-12 gives an example where three different threshold levels have been used.

Thus the technique has a number of required properties if the scheme is to serve the needed role for a sample classifier. However, several problems remain. Chief among these is that closed boundaries are not guaranteed, and often do not occur. Image processing algorithms

---

[101] A. G. Wacker, and D. A. Landgrebe, Boundaries in multispectral imagery by clustering, *IEEE Symposium on Adaptive Processes*, University of Texas at Austin, December 7-9, 1970.

to form closed boundaries from these would require substantial additional computation.

At this point in the development, the thought occurred that perhaps instead of seeking boundaries in the data, i.e., points where a field or object stops, one should look for fields or objects directly. That is, one should look for regions that are statistically homogeneous. The boundaries would then be determined by the edges of the objects. This approach would thus solve the problem of guaranteeing closed boundaries, and it seemed more logically related to the goal, that of identifying areas that are statistically homogeneous and can then be classified using a sample classifier.

Figure 8-11. Field boundqries by clustering on real data.

A possible approach to this end is the so-called "disjunctive" approach. The

Figure 8-12. Field boundaries with varied thresholds.

idea is to successively subdivide the image by quarters and test each quarter for a characteristic regularity as measured by an appropriately defined criterion function. For example, in a Landsat MSS image, urban areas are expected to have a unique type of texture. If this texture is not found to be occurring over the entire subimage being tested, the subimage is again quartered and each of the resulting subimages tested again. The process is repeated until the regularity criterion is satisfied. An example of this process is shown in the Figure 8-13.[102]

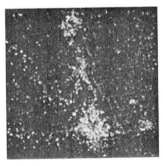 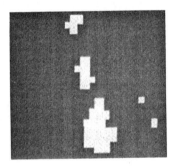

Landsat MSS Channel 2   Urban areas by disjunctive means.

Figure 8-13. Finding objects in data.

Though this method does provide the needed closed boundaries, it appears to provide boundaries that are somewhat crude.

[102] T. V. Robertson, K. S. Fu, and P. H. Swain, Multispectral image partitioning, LARS Information Note 071773, Laboratory for Applications of Remote Sensing, Purdue University, West Lafayette, IN, 1973.

[103] P. E. Anuta, E. M. Rodd, R. E. Jensen, and P. R. Tobias, Final report of the LARS/Purdue-IBM Houston Science Center joint study program, LARS Information Note 043072, Laboratory for Applications of Remote Sensing, Purdue University, West Lafayette, IN, 1972.

An alternate approach to object finding might be called a conjunctive approach. It would be to start with 2 x 2 pixel groups and grow a field by testing neighboring groups for commonalty with a standard statistical test. Such a process was defined[103] and coupled with a sample classifier to form the preliminary version of a complete analysis algorithm called "BOFIAC" for Boundary Finding and Classification.[104] The sample classifier used a statistical separability measure calculated from the mean vector and covariance matrix of the pixels in the object found. If the object contained fewer than the number of pixels needed for valid estimates of the covariance matrix at the dimensionality used, then the pixels of the object were classified individually with a standard maximum likelihood pixel classifier.

Example results are shown for two flightlines in Figure 8-14. The form of the output is in alphanumeric maps whereby the edges of fields are marked with letters designating which class the field had been classified to. The CBWE Flightline 210 data set was collected on August 13 in northern Indiana at an altitude of 5000 feet. Bands 0.61–0.70 μm, 0.72–0.92 μm, and 9.30–11.70 μm were used, based on minimum transformed divergence from the 12 bands available. The graph of Figure 8-15 (left) shows an accuracy comparison with conventional pixel classification for a 7253 pixel test set.

The Flightline C1 test data was flown at 2600 feet altitude above terrain on June 28. Bands 0.40–0.44 μm, 0.66–0.72 μm, and 0.80–1.00 μm were used for the classifications and a 19,246 pixel test set for the eight-class classifications. The graph of Figure 8-15 (right) shows the comparative results.

---

[104] R. L. Kettig and D. A. Landgrebe, Automatic boundary finding and sample classification of remotely sensed multispectral data, LARS Information Note 041773, Laboratory for Applications of Remote Sensing, Purdue University, West Lafayette, IN, 1973.

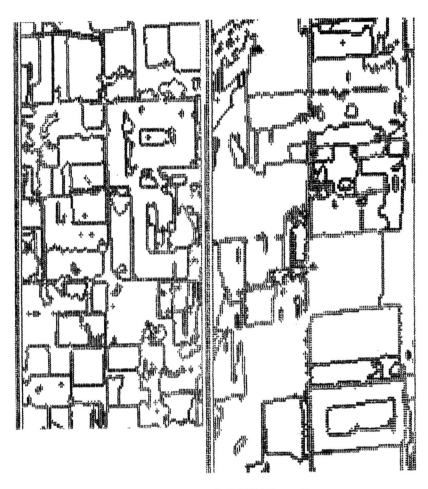

Figure 8-14. Example BOFIAC results.

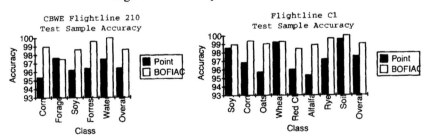

Figure 8-15. Comparison of BOFIAC results a pixel classifier.

8.8 The Design of the ECHO Classifier[105]

These tests appeared promising for the concept of spectral/spatial classification, and it seemed enough had been learned about spatial information in multispectral image data to define an effective spectral/spatial classifier for multispectral data that would make good use of spatial variations in the data. The basic scheme would need to involve two basic parts, scene segmentation into objects, followed by a sample classification algorithm in which each object delineated would be classified based on the statistical properties of the object's pixels.

Scene Segmentation. The concepts for defining the object boundaries studied had included both

- Boundary Seeking.
- Object Seeking

methods. Of the object seeking approaches, two types had been examined:

- *Conjunctive* methods which begin with a very fine partition and merge adjacent elements together to form homogeneous objects.
- *Disjunctive* methods which begin with a very simple partition and subdivide the partitions until each element satisfies the desired statistical homogeneity criterion.

---

[105] R. L. Kettig and D. A. Landgrebe, Classification of multispectral data by extraction and classification of homogeneous objects, *IEEE Transactions on Geoscience Electronics,* Vol. GE-14, No. 1, pp 19-26, January 1976.

It appeared that the conjunctive, object-seeking approach would be best. The final form for the scene segmentation portion of the algorithm thus selected was to be

A Two Stage Conjunctive, Object Seeking Approach involving
  1. Cell Selection
  2. Annexation

These steps are described next. Let the diagram of Figure 8-16 symbolize a portion of a multispectral image in which the small squares indicate individual pixels. Define <u>cells</u> to be of size n x n pixels, e.g., 2 x 2 pixels as shown by the heavier rectilinear lines. Then the two step process would be first to test a cell with a mild test of statistical homogeneity. If it passes, then test its neighbor on the right. If it passes also, use a statistical test to determine if the two have common statistics. If so, the second cell is said to be annexed to the first. Proceed to the next cell and test it for homogeneity and annexation. In this way, a region, referred to as an <u>object</u>, having common statistical characteristics continues to grow across and down until one or the other of the tests fails,

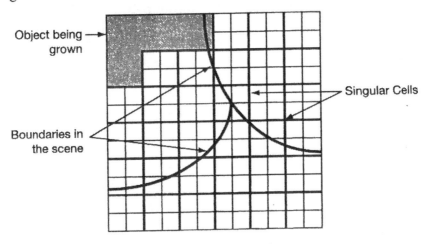

Figure 8-16. The ECHO field finding scheme.

405

indicating a boundary in the scene. Cells that fail the initial homogeneity test would be called singular and would be dealt with on a pixel by pixel basis.

Two modes for the scene segmentation were defined, *supervised*, in which knowledge of the class-conditional densities would be used, and *unsupervised*, in which they would not. The actual criteria used in the two cases are as follows.

Cell Selection

Unsupervised   Supervised

$$\dfrac{\sqrt{\text{sample variance}}}{\text{sample mean}} \begin{array}{c} > \\ < \end{array} \gamma \begin{array}{c} \to \text{Reject} \\ \to \text{Accept} \end{array}$$

$Q_J = (x_L - \mu_J)^T \Sigma_J^{-1}(x_L - \mu_J)$

$J = \text{Index} [\text{Max} \{p(x|K)\}]$

Annexation

Current (Partial) Object: $X = \{x_1, x_2, \dots x_N\}$

New Cell Under Test: $Y = \{y_1, y_2, \dots y_M\}$

F = Underlying density function of sample X

G = Underlying density function of sample Y

Generalized Likelihood Ratio Test:

$$\Lambda = \dfrac{\underset{\substack{F \in \Omega \\ F = G}}{Max\, p(\mathbf{x},\mathbf{y}\,|\,F,G)}}{\underset{\substack{F \in \Omega \\ G \in \Omega}}{Max\, p(\mathbf{x},\mathbf{y}\,|\,F,G)}} \begin{array}{c} > \\ < \end{array} T \begin{array}{c} \to Reject \\ \to Accept \end{array}$$

Unsupervised   Supervised

$\Omega = \{$Multivariate $\Omega = \{p(X|J): J = 1, 2, \dots, K\}$

Normal Density Functions$\}$

At the outset, it was assumed that the supervised approach, being more directly focused upon the data classes at hand, would provide the most useful results. The unsupervised method was carried along because, at a minimum, it might provide a useful tool for processing tasks when only scene segmentation was desired, classification was not desired, and thus no training samples were available.

8.9 Sample Classification

For the classification portion of the algorithm, any form of sample classifier might be a candidate. A statistical separability metric might be used, except that it fails for small objects since the statistical separability parameters cannot be satisfactorily estimated. Instead, a new scheme for sample classification, the maximum likelihood sample classifier, is defined by analogy to the maximum likelihood pixel classifier, as follows:

For the pixel classifier, the maximum likelihood strategy states

$$\text{Choose the class J by} \quad \underset{J}{Max} \; p(\mathbf{x}|J)$$

For the sample classification, the sample is now a collection of points $\mathbf{X} = \{\mathbf{x}_1, \mathbf{x}_2, ..., \mathbf{x}_N\}$ and the maximum likelihood strategy becomes

$$\text{Choose class J by} \quad \underset{J}{Max} \; p(\mathbf{x}_1, \mathbf{x}_2, ... , \mathbf{x}_N|J)$$

This might, at first glance, appear to an unreasonable strategy, since objects would no doubt have different numbers of pixels, N, necessitating estimating a large number of different density functions. This problem is avoided by a key simplifying assumption, making this more tractable. Assume that the pixels are class conditionally uncorrelated, and thus

$$p(\mathbf{x}_1,\mathbf{x}_2,\ldots,\mathbf{x}_N|J\} = \prod_{k=1}^{N} p(\mathbf{x}_k \mid J)$$

Again, the validity of this simplifying assumption might at first seem questionable, since it is known that pixels in multispectral image data usually have significant spatial correlation. However, it is noted in this case that all pixels under consideration are within an object, and thus are members of the same class. Tests on data show that for this class-conditional case, the correlation ordinarily decreases very rapidly with distance, making the assumption much more reasonable. As an illustration of this, Figure 8-17 shows the correlation vs. pixel separation for a data set and for two classes within that data set.

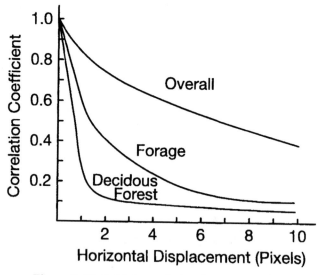

Figure 8-17. Spatial correlation for various classes.

Initial tests of the new algorithm, called ECHO for Extraction and Classification of Homogeneous Objects, appeared favorable. Figure 8-18 gives a comparison of the classification results from the pixel classifier and ECHO with identical training statistics on a moderately difficult classification task.

408

Pixel Result   ECHO Result

Figure 8-18. Thematic classification maps. (In color on CD)

Figure 8-19 shows quantitative evaluations in which special cases of parameter settings are used to determine which aspects of the ECHO algorithm seem to be the more important in improving accuracy.

The first bar in each graph shows the result for the per point classifier for reference. For the second, cell selection only, the annexation threshold was set so that there would be no annexation. Thus sample classification was being done on each 2 x 2 nonsingular cell. The third bar shows the result for the unsupervised case with the best setting of the annexation threshold. The fourth bar is the primary result. These latter two indicate the gain realized in growing objects from the cells, and that, as expected, the supervised mode outperforms the unsupervised mode. The last two bars are per field classifications, again for reference. In this case the boundaries were determined manually and either the maximum likelihood sample classification scheme or a statistical separability scheme was used. This case would show up the impact of any imperfect segmentation that occurred, since only the test fields, defined manually were classified. An important aspect of the new scheme is the degree of

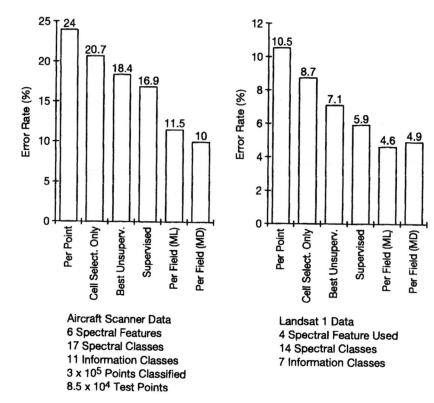

Aircraft Scanner Data
6 Spectral Features
17 Spectral Classes
11 Information Classes
3×10^5 Points Classified
8.5×10^4 Test Points

Landsat 1 Data
4 Spectral Feature Used
14 Spectral Classes
7 Information Classes

Figure 8-19. Test results for various parts of ECHO.

sensitivity of the results to the annexation threshold setting. Figure 8-20 at right shows the results of an evaluation of this characteristic on a Landsat 1 data set, and indicates that the threshold setting is not very sensitive.

ECHO was also tested against the more traditional kinds of texture measures such as those discussed above.[106] Figure 8-21 shows the

---

[106] D. J. Wiersma, and D. A. Landgrebe, The use of spatial characteristics for the Improvement of multispectral classification of remotely sensed data, *Proceedings of the Symposium on Machine Processing of Remotely Sensed Data*, IEEE Catalog No. 76CH1103-1 MPRSD, West Lafayette, IN, June 1976.

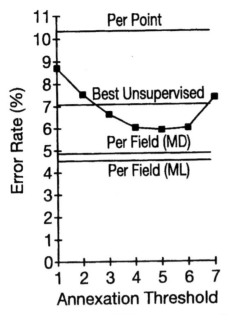

Figure 8-20. ECHO parameter sensitivity.

results of such a test. In this case the data was from Landsat MSS. Texture features were computed from channel 4 with a block size of 11. The optimum set consisted of the first two spectral bands

Figure 8-22 shows a comparison for different block sizes of the texture measures and utilizing different spectral bands from which the texture features were computed. For REF and ECHO only the four spectral bands were used. All spectral and texture features were used for the remaining classifications. The results indicate again that the pixel classifier gives the poorest results, that the texture feature scheme gives intermediate results no matter the band or block size selected, and the ECHO gives the best results

And finally, a test was run on various combinations of MSS spectral only and spectral plus one to four texture features. The four texture features used were Angular Second Moment, Contrast, Correlation,

411

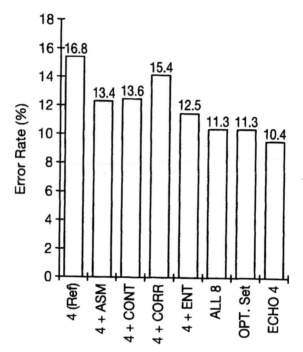

Figure 8-21. ECHO with texture measures.

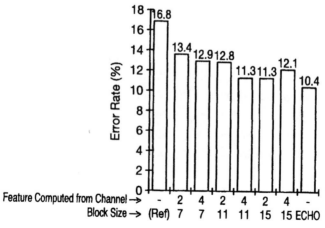

Figure 8-22. ECHO test results.

and Entropy as determined from band 2 of the MSS data with a block size of 15. The results are as shown in Figure 8-23. These results suggest that the marginal advantage of an added texture feature is considerable, that sample classification provides a somewhat greater advantage, but that an added texture feature used with sample classification is still better. Further, there appears to be no advantage in adding different types of texture features.

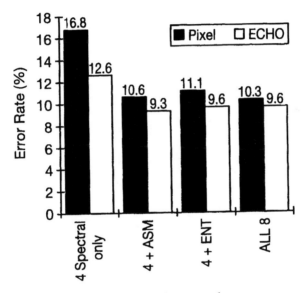

Figure 8-23. ECHO test results.

8.10 Test of the Algorithm

Given these favorable preliminary performance test results, next will be described a major test program carried out to more

107 Kast, J. L. and B. J. Davis, "Test of Spectral/Spatial Classifier," Final Report, NASA Johnson Space Center, Contract NAS9-14970, Vol. III, November 1977.

completely define the performance of ECHO.[107] The purpose of the test program was to,

- Verify the performance characteristics observed in the preliminary tests,
- Provide performance illustrations to the user community, and
- Determine rules for the choice of classifier parameters.

The test program included definition of a more complete set of performance indicators:

1. TRAining sample classification accuracy (TRA)

2. Field Center Pixel classification accuracy (FCP)

3. Full Field pixel classification Accuracy (FFA)

4. Root Mean Square proportion estimation error (RMS).

$$RMS = \sqrt{\frac{1}{N}\sum_{i=1}^{N}(C_i - C_i')^2}$$

C_i = % of pixels classified into class i
C_i' = % of region containing class i estimated from ground observation

5. Average classification VARiability (VAR)

$$VAR = \frac{(NCC)100}{50(NS-1)}$$

where NCC = number of class changes in 50 selected lines
NS = number of pixels per line.

6. Computer Processing Unit time (CPU).

The second and third of these are accuracy indications from test samples. FCP fields are selected staying sufficiently far back from field boundaries so as to ensure that no edge effect pixels are involved. FFA fields, on the other hand are selected to include as much of the field as possible, without obviously involving mixed, boundary pixels. The RMS measure is intended to determine if the correct number of pixels are being assigned to each class, which helps to determine if mixed boundary pixels are being properly accounted for. The fifth metric, VAR, is intended as a measure of the "salt-and-pepper" errors and to determine the extent that the classifier has smoothed out this characteristic.

Data sets assembled for the test included the following:

- 21 Data sets from Landsat MSS and two airborne scanners.
- IFOV's of 3, 6, 30, 40, 50, 60, and 80 meters.
- Sites in Indiana, Illinois, Kansas, and North Dakota.
- Collection Dates from April to September.
- Total pixels exceeding 10^6.

The intent was to have an appropriately broad range of spatial resolutions, of different landscape types, and different times of the season. Test were conducted for a large number of parameter values, including

- 768 observations of supervised ECHO.
- 81 observations of unsupervised ECHO

on the Landsat data sets alone, where the term "observation" is meant the classification of one data set with one set of parameters. The gray scale classifications of Figure 8-24 show a sampling of the results in pictorial form.

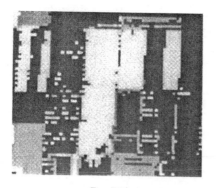
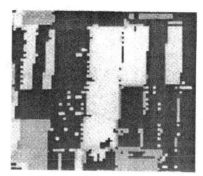

	Per Point		Low Threshold	
87.5%	Field Center Pixel Performance		90.0%
7.3%	RMS Proportion Error		2.5%
1200 Sec.	Computation Time		754 Sec.
34.8%	Classification Variability		29.9%

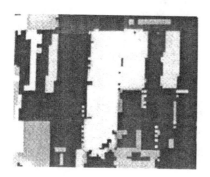
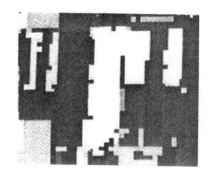

	Medium Threshold		High Homogeneity Threshold	
91.4%%	Field Center Pixel Performance		90.7%
2.5%	RMS Proportion Error		2.8%
494 Sec.	Computation Time		370 Sec.
22.5%	Classification Variability		18.6%

Figure 8-24. ECHO test comparison results.

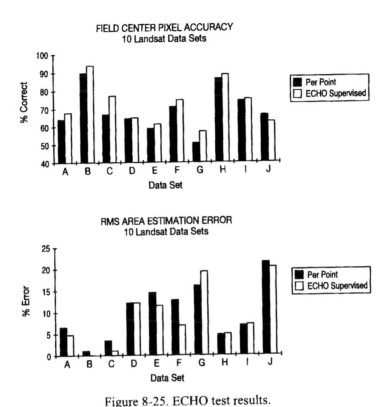

Figure 8-25. ECHO test results.

Figure 8-25 shows a sampling of the results in quantitative form.

The results were evaluated using analysis of variance tests. From this the following conclusions were reached.

- Supervised ECHO TRA, FCP, and FFA results are significantly superior to pixel and unsupervised ECHO.
- Unsupervised ECHO FFA are significantly inferior to the pixel results.
- Both ECHO processors gave significantly lower VAR and CPU values.

417

9 Noise in Remote Sensing Systems

In this chapter the subject of noise is taken up, extending the information begun in Chapter 2. Noise must be treated in relation to other parameters, as its effect varies depending on other system circumstances. Thus a system model relative to noise is presented and its impact evaluated. A more elaborate model for both signal and noise is then presented and example effects are explored.

9.1 Introduction

Earlier, it was pointed out that the spatial sampling scheme, the spectral sampling scheme and the signal-to-noise ratio are the three fundamental parameters that define the information content of the data gathered by the sensor. In this chapter the matter of noise and the signal-to-noise ratio is explored in additional detail.

As should be apparent from the preceding discussion, noise and randomness are not the same thing. The random nature of a signal does not imply that it is simply noisy. Indeed, randomness often has a kind of structure about it that can be quite information bearing but may not be apparent by simple visual inspection. This makes it difficult to state a strict, simple definition of noise and signal-to-noise ratio. A useful though subjective definition of the difference between signal and noise may be stated as follows:
- Signal is any signal variation that is information-bearing
- Noise is any signal variation that is not

Thus the definition of signal and noise is problem dependent. In one instance, a signal variation may be "signal," and the same variation in the same data set but for a different application can be "noise." For example, suppose that one has spectral data over an agricultural area where there is only a partial crop canopy such that a pixel contains components from both the vegetation and the soil. Then to the soil scientist, the spectral variation from the vegetation is "noise," whereas to the crop scientist, the spectral variation emanating from the soil is "noise."

Generally, once noise from whatever the source is injected into the data stream, it is very difficult, if not impossible to remove. This is because the information about the component of the signal that is the noise is often not known or not known to sufficient precision, such that any attempt to use it results itself in the injection of a new source of noise. On the other hand, there are components of noise that can clearly be identified as such based on their point of origin. Random variations created within the sensor system itself must surely be regarded as noise if the data stream is intended to provide information about the ground scene. It is this latter kind of noise that will serve as the focus for the discussion for the remainder of this chapter. The discussion will be begun by some simple examples of the effect of noise on the discrimination process.

9.2 Example: The Effects of Noise

An experiment that aids in showing the relative impact of noise is first described. In this case a study was carried out to determine the effects of noise on pattern recognition analysis of multispectral data.[108] The data were gathered from an agricultural area by a multispectral scanner, then simulated noise in the form of white Gaussian random numbers in varying levels was added to the data

---

[108] The results displayed are a recent reconstruction of the experimental results which originally appeared in P. J. Ready,, P. A. Wintz, S. J. Witsitt and D. A. Landgrebe, Effects of compression and random noise on multispectral data, *Proceedings of Seventh International Symposium on Remote Sensing of Environment*, Environmental Research Institute of Michigan, Ann Arbor, pp. 1321–1343, 1971.

to form modified data sets. Identical analyses were carried out on each and the results compared. Figure 9-1 below shows in image from a small segment of one channel of the data with low, moderate, and high levels of noise added. The noise level is denoted by σ, the standard deviation of the noise in units of digital count, i.e., $\sigma = 5$ signifies a noise level in which the standard deviation of the noise added was equal to five brightness levels out of the total brightness-level range, which was 256 in this case. The mean value of the

$\sigma = 2$

$\sigma = 7$

$\sigma = 15$

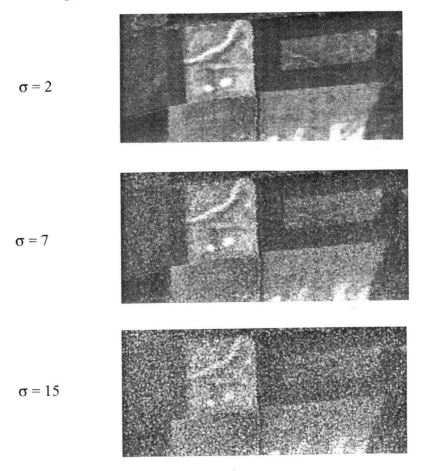

Figure 9-1. Examples in image form of noise added to data.

421

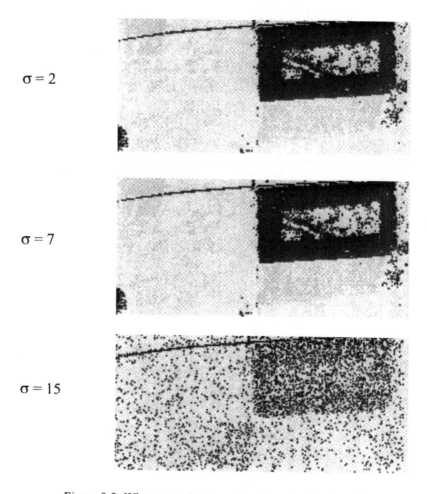

$\sigma = 2$

$\sigma = 7$

$\sigma = 15$

Figure 9-2. Wheat vs. other for various levels of noise added.

noise added was 128 in every case. In addition to the original (no noise) data set, sets were generated for $\sigma = 2, 5, 7, 10, 15,$ and 17.

Classification of the full data set (220 samples per line and 949 lines) was carried out with the same four bands in each case (0.40–.44 μm, 0.52–0.55 μm, 0.66–0.72 μm, 0.80–1.00 μm). The classification was into 10 classes that were then grouped into eight groups for accuracy determination. For the 10 classes a total of

422

CLASSIFICATION PERFORMANCE VS. NOISE

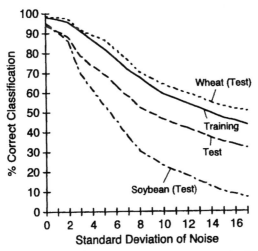

Figure 9-3. Classification accuracy vs. noise level.

1407 training samples were used, and a test set of 70,635 samples was established.

Figure 9-2 shows example results in thematic map form for the classWheat. Pixels classified as Wheat are shown in black; all others are in white. This allows one to perceive on a qualitative basis the manner in which the classification accuracy deteriorated as the noise level increased.

Figure 9-3 shows the results of the test in quantitative form. The overall average accuracy and the accuracy for two specific classes are presented. While the general shape of these curves shows the expected downward trend, a significant point to be noted here is with regard to the two individual classes Wheat and Soybeans. The data set was gathered in early summer over a test site in the U.S. Corn Belt. At that time of year, wheat has ripened to a golden brown and was ready for harvest. Soybeans, on the other hand, having been planted only a few weeks before, were green but had not developed a full canopy, and the percentage of ground cover was still rather small. As a result the Wheat class is relatively easy to distinguish from the other classes present in the scene, while the Soybean class represents a much more difficult discrimination problem. Notice that, although both began with high accuracy, above 90%, as noise is added to the data, the accuracy of the difficult class, Soybeans, deteriorates much more rapidly than that of Wheat. This suggests that a more marginally identifiable class would be more adversely affected by a deteriorating signal-to-noise ratio.

One might suspect that in feature space the effect of the noise is to broaden the distribution of each class. Thus a class that is easily distinguishable, such as Wheat, which has no other class near it in feature space, would not be hurt much. A class like Soybeans that is more difficult to discriminate from others would be affected more strongly, since the effect of the noise broadening its distribution would bring it into conflict with other classes nearby in feature space.

9.3 A Noise Model[109]

Although there are many noise sources, the study of different kinds of noise will begin here by considering just three sources of noise.

- Atmospheric Effects

    - We will assume that the atmosphere is uniform and constant over the area to be studied,[110] and shall account

---

[109] D.A. Landgrebe and E.R. Malaret, "Noise in Remote Sensing Systems: Effect on Classification Accuracy," *IEEE Transactions on Geoscience and Remote Sensing*, Vol. GE-24, No. 2, pp. 294-299, March 1986

only for average transmissivity and path radiance at this time.

- Detector/Preamplifier Noise Processes

  - Shot noise, noise due to the random time of creation and recombination of charge carriers in the semiconductor detector.

  - Thermal noise, noise due to the random components of motion of electrons in a resistance that is above absolute zero temperature. It is white, Gaussian, zero mean, and with variance $\overline{v^2} = 4\,kTR\Delta f$ where k is Boltzman's constant, T is absolute temperature, R is the resistance and Δf is the bandwidth.

- Quantization Noise

  - The error that is introduced when the continuously varying signal is discretized in the process of A/D conversion.

In particular, the model of Figure 9-4 will be used to study the effect of noise on such information systems. In this model, the following are assumed.

- A linear model, $L(\lambda) = \tau_a\, L_s(\lambda) + L_p(\lambda)$, is used for the atmospheric effect.

- Shot noise, generated in the sensor detector, is zero mean, Gaussian, with variance directly related to the signal level.

---

110 Clearly this is an overly simplified assumption.

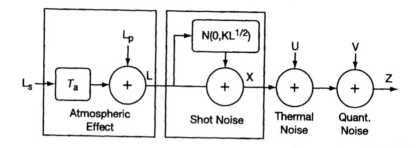

Figure 9-4. System Noise Model.

- Thermal noise is zero mean, Gaussian noise with variance proportional to the temperature and resistance of the amplifier input and the bandwidth of the amplifier.

- Quantization noise is uniformly distributed over one discrete quantization interval.

First, we make the following observation. The role that the atmospheric effect has on data analysis is one that is oftentimes misunderstood. For this linear model of the atmospheric effect, the atmosphere should have *no* effect on the separability of classes, since a linear transformation on a feature space does not change the separability of the classes in it. To illustrate in a specific case, assume that the classes are multivariate normal, and denote the density function for class j before the radiance passes through the atmosphere as

$$p(\mathbf{L_s}|\omega_j) = N(\mu_j, \Sigma_j)$$

that is, a multivariate Gaussian density function with mean vector μ_j, and covariance matrix Σ_j. Then if τ_a is a diagonal matrix of the atmospheric transmissivity, the density function for the radiance after passing through the atmosphere would be

426

$$p(L|\omega_j) = N(\tau_a\mu_j + L_p, \tau_a\Sigma_j\tau_a^t)$$

That is, the mean value after passing though the atmosphere would be $\tau_a\mu_j + L_p$ and the covariance matrix would be $\tau_a\Sigma_j\tau_a^t$. From this and the previous expression, it is easy to show that,

$$p(\mathbf{L}|\omega_i) = \frac{1}{|\tau_a|}p(\mathbf{L}_s|\omega_i)$$

Since the transformed density function differs from its original density by only a multiplicative constant, and since the same multiplicative constant would apply to all classes, the classification statistics in a multiclass problem would be unaffected. It can also be shown that the correlation coefficients between channels are preserved by the linear transformation performed by the atmosphere. Obviously the atmosphere does have a degrading effect on analysis results, but not by itself. The degradation occurs in concert with other effects, as will shortly be seen.[111]

Since the thermal, shot, and quantization noises are zero mean, it is seen that the mean vector of the output would be given by

$$E[Z] = E[L] = \tau_a E[L_s] + L_p$$

The output covariance matrix may be determined to be

---

[111] A more complete discussion of this point is given in J. Hoffbeck and D. A. Landgrebe, Effect of radiance-to-reflectance transformation and atmosphere removal on maximum likelihood classification accuracy of high-dimensional remote sensing data, *Proceedings of the International Geoscience and Remote Sensing Symposium* (IGARSS'94), CD-ROM pp 3289–3294, Pasadena, Calif, August 8–12, 1994.

$$\Sigma_z = \begin{bmatrix} \sigma_{L_1}^2 + k_1^2 \bar{L}_1 + \sigma_{U_1}^2 + \sigma_{V_1}^2 & & \\ \rho_{L_1 L_2} \sigma_{L_1} \sigma_{L_2} & \sigma_{L_2}^2 + k_2^2 \bar{L}_2 + \sigma_{U_2}^2 + \sigma_{V_2}^2 & \\ \cdot & & \\ \rho_{L_1 L_n} \sigma_{L_1} \sigma_{L_n} & \rho_{L_1 L_n} \sigma_{L_1} \sigma_{L_n} & \sigma_{L_n}^2 + k_n^2 \bar{L}_n + \sigma_{U_n}^2 + \sigma_{V_n}^2 \\ & & \cdot \end{bmatrix}$$

The correlation coefficient between two bands at the output may be shown to be

$$\rho_{z_i z_j} = \rho_{L_i L_j} \left[\frac{\sigma_{L_i} \sigma_{L_j}}{\sqrt{\left(\sigma_{L_i}^2 + k_i^2 \bar{L}_i + \sigma_{U_i}^2 + \sigma_{V_i}^2 \right) \left(\sigma_{L_j}^2 + k_j^2 \bar{L}_j + \sigma_{U_j}^2 + \sigma_{V_j}^2 \right)}} \right]$$

It is easy to show that the correlation between input and output for a given spectral band is given by

$$\rho_{Z_i L_{si}} = \frac{\tau_a \sigma_{L_{si}}}{\sqrt{\tau_a^2 \sigma_{L_{si}}^2 + k_i^2 \left(\tau_a \bar{L}_{si} + L_p \right) + \sigma_{U_i}^2 + \sigma_{V_i}^2}}$$

As a test of consistency, note the following.:

- For zero atmospheric transmittance, this correlation coefficient is zero,

- For no system noise, $k_i = \sigma_{Ui} = \sigma_{Vi} = 0$, the input-output correlation coefficient is unity. Note that this is true regardless of the condition of the atmospheric variables, τ_a and L_p. This again implies that without other sources of noise, the atmosphere has no deleterious effect.

428

- As the path radiance increases, the input-output correlation coefficient decreases, indicating loss of information.

9.4 A Further Example of Noise Effects

To illustrate the relative effects of these noise sources, example settings for the model parameters are used. These settings approximate those of the Thematic Mapper, but for data from the Michigan M-7 aircraft scanner[112] operated over a Corn Blight Watch Experiment flightline in northern Indiana. The magnitude of the noise components and their combination are shown in Figures 9-5, 9-6, and 9-7. Classification accuracies that would be obtained are determined as shown in Figures 9-8.

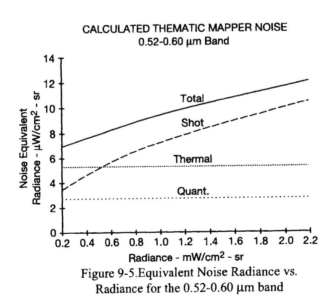

CALCULATED THEMATIC MAPPER NOISE
0.52-0.60 µm Band

Figure 9-5.Equivalent Noise Radiance vs.
Radiance for the 0.52-0.60 µm band

---

112 See P.H. Swain and S.M Davis, eds. *Remote sensing: the quantitative approach*, McGraw-Hill 1978, p.122.

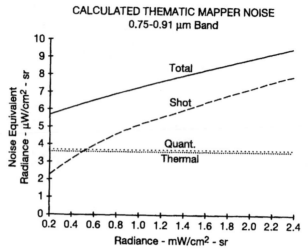

Figure 9-6. Equivalent Noise Radiance vs.
Radiance for the 0.75-0.91 μm band.

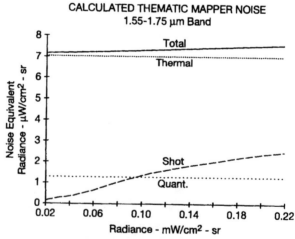

Figure 9-7. Equivalent Noise Radiance vs.
Radiance for the 1.55-1.75 μm band.

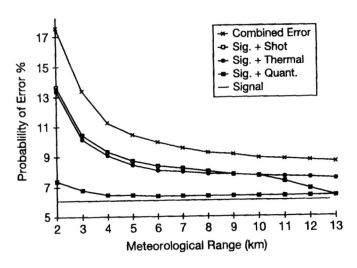

Figure 9-8. Probability of error vs. meteorological range.

From Figure 9-8, it is seen that for the signal only case (with only atmospheric effects and without shot, thermal or quantization noise), the error is unaffected by the atmosphere. Even for the Signal + Quanttization case, the probability of error of the curve is relatively flat with only a slight rise when the meteorological range becomes as poor as 2 km. Adding any other noise sources quite significantly affects the error, especially at low meteorological range. Notice also that for the case of Signal + Shot Noise, the performance is similar to that with Signal + Thermal Noise with improving atmosphere, up to a point (Met. range ≈ 10 km). Beyond this point, the Signal + Shot Noise begins to increasingly approximate the signal only performance. This may be due to the fact that for lower meteological range, the increased radiance due to path radiance, radiance unrelated to the target, tends to increase the shot noise causing the performance to deteriorate beyond that for the signal alone.

Figure 9-9 shows probability of error vs. quantization. Notice that in this case the curve becomes fairly flat after 7 bits, indicating that

431

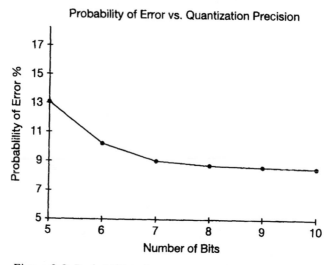

Figure 9-9. Probability of error vs. quantization precision.

for the particular parameter settings used in this case, more than 128 gray levels did not improve the accuracy. However, this result is closely related to the dynamic range of the scene, and as such, the 7 bits quantization is probably the minimum number of bits that would be needed for the optimal conditions of scene and sensor parameter settings. This illustrates the value of this kind of simulation. One could experiment with other parameter settings, for example, to develop perceptions as to which parameter values have the most sensitive relation to system performance.

9.5 A Signal and Noise Simulator[113-115]

As has been seen, there are many factors that affect the information content and the ability to extract information from the electromagnetic field arising from the Earth's surface. The scientific and engineering aspects of these factors have been studied extensively. However, most of these studies have been aimed at understanding effects of individual noise sources on the remote sensing process, while relatively few have studied their

432

interrelations. In the previous section, the interactions of four of these was studied, however there are many more.

To illustrate, Figure 8-10 shows a taxonomy of components and effects that can degrade the information extraction process. In a similar manner, a comparable taxonomy may be developed for signal variations and states that contribute to the potential information output of the system. Figure 8-11 is an example signal taxonomy of such effects. These taxonomies offer a framework in which remote sensing system effects can be grouped and located. The categories under the main subsystems delineate sources of major contributions to the system state. In some cases, effects or sources are listed in both signal and noise structures. These dual listings exemplify one of the major problems in understanding remote sensing systems. Depending on what type of information is desired, sources or effects may indeed represent both noise and signal effects.

One way to study the possible interactions of such a complex set of factors is to construct a stimulator of an entire remote sensing system so that information extraction results under various circumstances and parametric value settings can be compared. Such a simulation scheme will be described next, with example results of its use given.

---

[113] J. P. Kerekes and D. A. Landgrebe, Simulation of optical remote sensing systems, *IEEE Transactions on Geoscience and Remote Sensing*, Vol. 27, No. 6, pp 762-771, November 1989

[114] J. P. Kerekes and D. Landgrebe, Parameter tradeoffs for imaging spectroscopy systems, *IEEE Transactions on Geoscience and Remote Sensing*, Vol. 29, No. 1, pp 57-65, January 1991.

[115] J. P. Kerekes and David Landgrebe, An analytical model of Earth observational remote sensing systems, *IEEE Transactions on Systems, Man, and Cybernetics*, Vol. 21, No. 1 pp 125-133, January 1991.

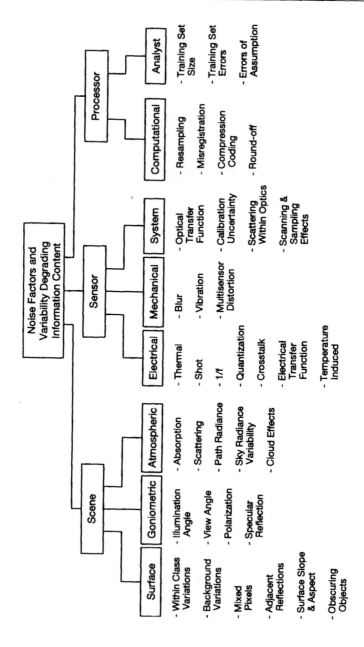

Figure 9-10. Noise Sources.

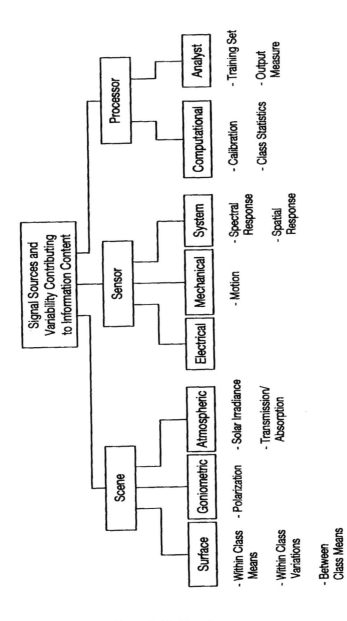

Figure 9-11. Signal sources.

Such a system simulator was constructed[116] and its overall organization is shown in Figure 9-12. The overall model is broken into three parts: the scene model, the sensor model, and the processing model. In each case, descriptive information of the model is fed in, defining what set of model parameter values are of specific interest. Each of the three subsystem models produces an output to be used by the next part of the system. The scene model produces a high spatial and spectral resolution data file of what the sensor will "see." The sensor model uses this data file to produce its output, which are data of the type to be analyzed by the processing model. The output of the processing model is the accuracy of the classification that would result from the specific scene and sensor parameter values and the classes and analysis algorithms specified for the processing model.

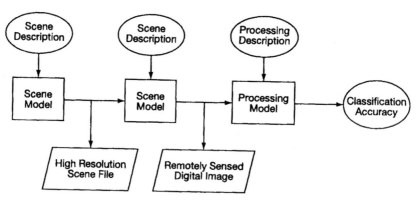

Figure 9-12. Remote sensing system model.

Figure 9-13 shows the Scene Model in greater detail. The scene model uses a two-dimensional autoregressive spatial model, Lambertian surface reflectance, and spectral characteristics derived

---

[116] J. P. Kerekes and D. A. Landgrebe, Simulation of optical remote sensing systems, *IEEE Transactions on Geoscience and Remote Sensing*, Vol. 27, No. 6, pp. 762-771, November 1989

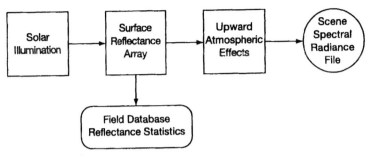

Figure 9-13. Scene model.

on a class-conditional basis from a database containing hundreds of thousand of spectra gathered in carefully controlled laboratory and field measurements. Table 9-1 shows some of the parameters whose values may be controlled as a part of a simulation. The conditions that can be simulated are limited by what data are available from the laboratory and field database. For example, there were no data available in the database to simulate the non-lambertian characteristics of many ground cover types.

Scene Parameters	Symbol
Surface meteorological range	V_η
Solar zenith angle	θ_{solar}
Solar azimuth angle	ϕ_{solar}
View zenith angle	θ_{view}
View azimuth angle	ϕ_{view}

Table 9-1. Atmospheric and goniometric parameters.

Figure 9-14 provides greater detail of the Sensor Model. By this means various aircraft and spacecraft sensor systems could be simulated. Table 9-2 gives a list of the parameters of the sensor model that may be varied as a part of a simulation.

437

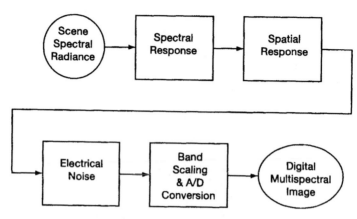

Figure 9-14. Sensor model.

<u>Sensor Parameters</u>	<u>Symbol</u>
Altitude (meters)	H
Cross-track line spread function	$h_x(\bullet)$
Along-track line spread function	$h_y(\bullet)$
Sum of $h_x(\ )$ coefficients	A_x
Sum of $h_y(\ )$ coefficients	A_y
Angular distance between $h_x()$ coefficients (radians)	ΔU
Angular distance between $h_y()$ coefficients (radians)	ΔV
Cross-track sampling interval (radians)	ΔW
Along-track sampling interval (radians)	ΔZ
Number of spectral bands	L
Band l spectral response	$s_l(m)$
Full scale band l radiance (mW/cm^2-sr)	$L_{Full,l}$
Shot noise standard deviation	$\sigma_s(l)$
Thermal noise standard deviation (mW/cm^2-sr)	$\sigma_t(l)$
Calibration error standard deviation	$\sigma_c(l)$
Number of radiometric bits	Q

Table 9-2. Sensor model parameters.

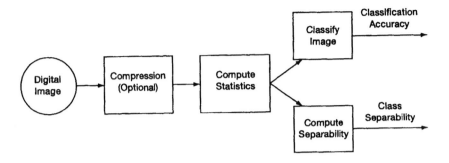

Figure 9-15. Block diagram of processing stage.

And finally, Figure 9-15 shows the organization of the Processing Model. From it one obtains either the classification accuracy for the particular classifier algorithm used or a quantitative measure of the class separability based on a selected statistical separability measure. These are regarded as the overall indices of performance for a given set of simulator parameter values.

Following are some results obtained in using the system simulator to indicate the effects of interrelationships among the many parameters. For this purpose a simulated scene was created for a Landsat 2 MSS sensing of fields containing four agricultural crop types. The spectral reflectance factor was then converted from the 20-nm wavelength spacing of the field instrument used to collect the data set to the 10 nm resolution used in the scene simulator through a simple interpolation algorithm.

Two demonstrations were performed to explore the role of goniometric effects on the classification accuracy. One simulating the sensor looking straight down and the sun angle varying, and the other with the sun at zenith and the sensor increasing its viewing angle. The results are shown in Figure 9-16. Note that these results were obtained while necessarily assuming Lambertian surface

439

reflectance characteristics. Since the surface reflectance did not change with the zenith angle in this experiment, the variation in the results most likely came from the effect of the atmosphere and geometry involved. The effect of the sun zenith angle increasing was a consistently decreasing classification accuracy, while the increasing view zenith angle shows a flat spot for view angles of 15°, 30°, and 45°. This flat spot is most likely due to two offsetting effects: as the view angle increases, there is a decreased signal radiance and increased path radiance resulting in a *lower* classification accuracy. But at the same time the number of high-resolution scene pixels within the sensor IFOV increases, thereby decreasing class variation and resulting in a *higher* classification accuracy.

Another experiment was performed to illustrate how the effect of the atmosphere relates to the sun angle. For this experiment the sensor was again the Landsat MSS at 0° zenith angle. Figure. 9-17 shows the result. While the atmosphere does indeed have a detrimental effect for all sun zenith angles, it can be seen that the effect is significantly more pronounced for lower sun angles. The

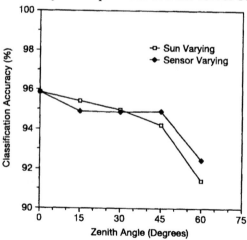

Figure 9-16. Goniometric effects.

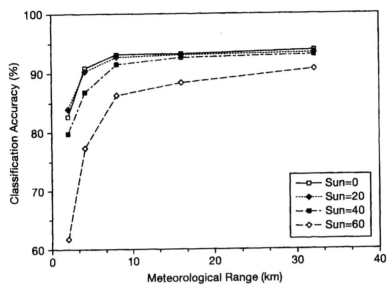

Figure 9-17. Effect of atmosphere for varying sun angles on MSS data

decrease in accuracy at low meteorological ranges is due to the reduced signal levels and increased shot noise due to path radiance, along with the high quantization error of the 6-bit radiometric resolution.

Next a demonstration is given to show how the radiometric resolution affected classification accuracy for various types of sensors. In addition to the Landsat MSS sensor used above, the Landsat TM and the SPOT MSS parameters were used. Figure 9-18 shows the results using a 10-km meteorological range. Here it appears that accuracy does not improve significantly at radiometric resolutions greater than 7 or 8 bits. This is due to the fact that above this resolution the other sources of error dominate the quantization error and limit the classification accuracy. In addition to the spatial and spectral band differences, the absolute level of classification accuracy for the various sensors may be related to the noise levels used in the simulation.

441

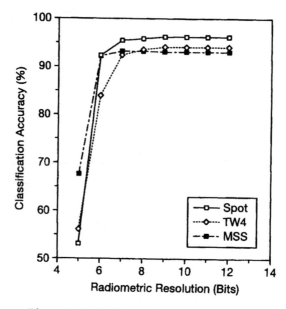

Figure 9-18. Radiometric resolution for various sensors
using a 10 km meteorological range

Again, the point of these results is to emphasize that it may be misleading to consider only one corrupting influence at a time. There are many such factors, and there is a significant tendency for their effects to be interrelated. Considering them one at a time may lead one to an inappropriate conclusion about how significant any one factor is.

10 Multispectral Image Data Preprocessing

This chapter treats a number of different types of preprocessing for multispectral data, including radiometric and geometric data adjustment, registration of data collected over the same area at different times or from different sensors, and goniometric effects that impact the utility of the data.

10.1 Introduction

Preprocessing generally precedes data analysis or information extraction as seen in Figure 10-1. Its goal is the reduction of distortion or the enhancement of some aspect of the data. Preprossing involves the intelligent use of ancillary knowledge or data about the scene or the collection process.

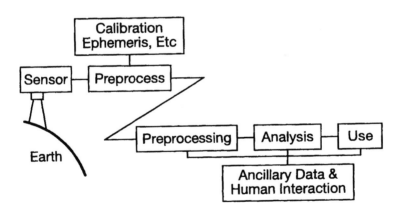

Figure 10-1. A system overview.

There are a very wide variety of types of preprocessing that are sometimes carried out on data. Some typical examples are:

- Radiometric preprocessing
  - Relative calibration
  - Absolute calibration
  - Removal of systematic distortion

- Geometric Preprocessing
  - Distortion reduction
  - Rectification
  - Image registration

- Data Transformations
  - Ratios
  - Edge enhancements
  - Feature extraction

The data transformation category was covered in an earlier chapter. Therefore the examples of this chapter will cover the first two categories.

10.2 Radiometric Preprocessing

Radiometric Calibration There are two types of radiometric calibration that are sometimes used as preprocessing steps: relative and absolute calibration. The intent of relative calibration is to make the data *internally* consistent, one measurement to another; absolute calibration is intended to make the data consistent with regard to an *external* standard, such as a system of geophysical units.

Frequently in sensor systems, the overall system gain will drift or slowly vary in some fashion, for example, due to power source variations or component aging. The effect makes a given power input into the sensor pupil produce different digital count values out at different times. The purpose of relative calibration is to minimize this effect so that the data after preprocessing is as if it had been produced by a sensor system whose overall radiometric gain was fixed at some constant but unknown value. Absolute calibration would take this one step further by adjusting that overall fixed gain to some known value, such as to provide an output in some specific geophysical units.

Approach

> Let x_{meas} = initial output for a given measurement, e.g. a given pixel response in a given spectral band.
>
> x_{rel} = value of this measurement after relative calibration.
>
> x_{abs} = value of this measurement after absolute calibration.

The general procedure for relative calibration is typically as follows:

1. Assume a linear relationship between x_{meas} and x_{rel}.

2. Two reference points are needed to establish the linear relationship, c_0, often at or near the black level or zero input, and c_1, often a value near maximum or full scale. See Figure 10-2. Sometimes a third calibration point is provided at an intermediate point, to continually verify the linearity of the system, the stability of the calibration, or for other purposes. In aircraft systems, calibration

lamps that are periodically passed into the field of view often provide these reference points. In spacecraft systems they may be established by looking at a particular point in the sky whose intensity is known to be stable, or perhaps such a location on the surface of the moon.

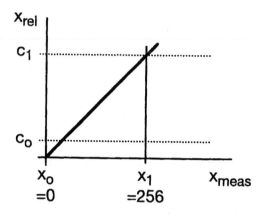

Figure 10-2. The linear calibration process.

Since the relationship between x_{meas} and x_{rel} is assumed linear,

$$x_{rel} = A\, x_{meas} + B$$

A and B are assumed to vary slowly and must be re-determined periodically. Since c_0 and c_1 are assumed to be fixed, they may be used to determine A and B as often as needed e.g., once per scan line. Thus,

$$c_0 = A\, x_0 + B$$
$$c_1 = A\, x_1 + B$$

If $x_0 = 0$ then $B = c_0$, and

$$A = \frac{c_1 - c_0}{x_1}$$

Thus

$$x_{rel} = \frac{c_1 - c_0}{x_1} x_{meas} + c_0$$

Note that if the actual radiometric values for c_0 and c_1 are known in absolute units, then the same procedure results in absolute calibration. Thus

- Relative calibration is intended to adjust for system drift, while

- Absolute calibration provides a more specific adjustment to actual radiometric units.

Though it appears straightforward, such a procedure can be problematic to carry out in practice. To begin to explore this, perhaps the simplest implementation of calibration occurred in the past days of the optical mechanical line scanner. As an example of this process in practice, consider the scheme used in the M-7 airborne scanner.[117] The arrangement of calibration sources for this system was as shown in Figure 10-3. Rotation of the scanning mirror caused the instantaneous field of view to be directed across the scene below and then, as the rotation continued, up inside the scanner body where the calibration sources and surfaces were located. Thus the calibration sources were viewed once each scan line and through the same optical path and the wavelength dispersive device to the detector for each wavelength band. This way each sensor element could view the calibration sources once for each scan line.

---

[117] See P.H. Swain and S.M Davis, eds. *Remote Sensing, The Quantitative Approach*, McGraw-Hill 1978, p.122.

Calibration sources consisted of one, two, or three calibration lamps and, for a time, a solar port with an attached fiber optic bundle leading to an opal dome on top of the aircraft, as well as thermally controlled surfaces for the thermal part of the spectrum. The solar observation was intended to add an additional element to the calibration process by attempting to adjust for variations in solar illumination on the ground scene pixels, but this part was not wholly successful. It was not possible to measure the illumination level at the ground pixel, only on the top of the aircraft. While this might allow for accounting for effects due to atmospheric conditions above the aircraft or changes in the solar output, one could do so only very generally and not on a pixel by pixel basis, since the illumination level falling on the aircraft would not necessarily be the same as that falling on any part of the scene being viewed.

Typical steps in the calibration process during data preprocessing for a reflective detector were as follows:

1. Sample and quantize each detector output at equal increments of time for the entire data collection run. Note that because of aircraft roll, the location of the calibration pulses would shift in the data record with respect to the location of the samples from the scene. As a result marker pulses were provided referenced to the frame of the sensor while roll-stabilized pulses were provided referenced to an inertially stabilized platform, and thus to the ground scene.

2. With a software-implemented search algorithm, locate relative to the marker pulse of each scan line the reference lamp response(s), the solar lamp response, and the dark level. Analogous techniques would be used for the thermal detector.

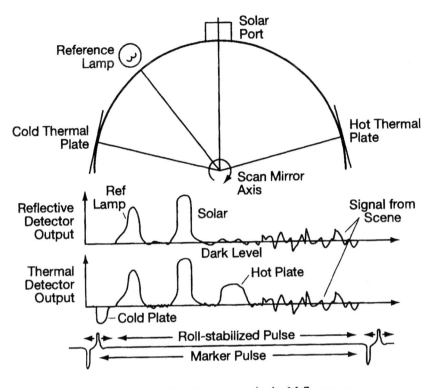

Figure 10-3. Calibration sources in the M-7 scanner

3. Determine the mean and standard deviation for these calibration source levels using a fixed number of samples about the centerline of these calibration pulses and the dark level.

4. Using the roll-stabilized pulse of each scan line, extract the appropriate number of samples of the scene for that scan line.

5. Make up a record for each scan line consisting of, in the case of the M-7 scanner, the 220 ground pixel samples

followed by the six calibration numbers, that is, the mean values of c_0, c_1, and c_2, and their estimated standard deviations. The latter three numbers were used for quality control assessment.

This process must be carried out for each reflective detector, with a similar process for the thermal band. Note that the calibration step itself is not carried out. Rather, the preprocessing only assembles the calibration data into an easily useful form. The calibration transformation, itself, is left for the discretion of the analyst during the application of the analysis algorithm. The reason for this is because of the problematic nature of the process. As a result of this, depending on the circumstances, one may or may not improve matters by attempting to calibrate the data, as will be seen shortly. Usually only relative calibration was used, since the relative differences in scene properties are enough to discriminate among cover types – the actual radiance values were usually not needed.

It is seen that calibration using the black level, c_0 in the above example and the near-full scale level, c_1 above, allows one to compensate for gain changes which may occur anywhere between the sensor detector and the digitizer, assuming that the calibration lamp brightness does not vary. However, this does not account for any variations that might take place in front of the sensor.

Figure 10-4 shows the results of using various degrees of calibration on three data sets. In these results, calibration using c_0 and c_1 appeared to give the best results, though not always. The explanation for this appears to be that (a) the solar output as measured at the top of the airplane was not necessarily the same as that irradiating the Earth's surface at the pixel currently being observed, and (b) the precision with which the calibration source values could be measured varied, introducing another source of noise. In practice,

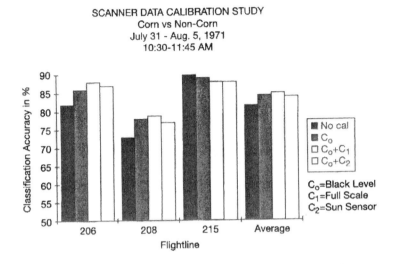

Figure 10-4. The effect of a calibration process in a realistic circumstance.

because of vibrating calibration lamp filaments and other such effects, the shapes of the calibration pulses were not usually as idealized as those shown in Figure 10-3, and varied significantly from scanline to scanline. That is why the variance of the samples used to estimate the values of c_0, c_1, and c_2 were recorded, as they tended to indicate when there had been a significant deviation in pulse shape.

A simple numerical example will serve to illustrate the point. Suppose that $x_{meas}= 100$, and

$$c_1 = 200 \pm 10 \quad x_1 = 220$$
$$c_0 = 10 \pm 3 \quad x_0 = 0$$

Then, using the calibration formula above–though the nominal value of $x_{rel} = 96.36$, given the uncertainty in the calibration data–x_{rel} can range over $90.18 \leq x_{rel} \leq 102.55$, an uncertainty range of 12% of nominal, and this ignores any uncertainty in x_{meas}. Thus much

451

of the success of calibration procedures hinge upon how accurately calibration data can obtained.

In practice it was found that, if for a given flight, there appeared to be high noise levels in the c_1 values, it sometimes proved better to calibrate using c_0 only. This was because, c_0 was a black level as compared to the output of a lamp filament, so the values measured from it tended to have lower variance.

In the early days, the signal-to-noise ratios for (aircraft) systems were such as to justify only 8-bit data systems, and this was seen to be a significant limitation on the extractable information content of data. Thus S/N was a target for improvement of sensor systems themselves. One way to accomplish this improvement was to increase the dwell time on pixels. For a fixed platform velocity, this could be done by scanning more than one scan line at a time. Thus for a given platform velocity, the integration time for power from a pixel would be multiplied by the number of scan lines being simultaneously scanned. For example, the Landsat MSS system scanned six lines at once. Improving the S/N of the sensor, tends to make the calibration process more difficult to carry out with the required precision, since now, instead of one sensor to calibrate per spectral band, there would be several. (See below.)

Further, as sensor system technology continued to develop over the years, the next step was to use two-dimensional arrays of detectors, one dimension for the various spectral bands, and the other for the row of pixels in the cross track direction. This had two effects on the calibration process. One was that there were now many more detectors that must be cross-calibrated, those in the cross-track direction. The other was that the dwell time on a pixel was increased substantially, since an entire row of pixels was being viewed simultaneously in parallel. As a result S/N ratios justifying

10- to 12-bit precision have become common. This imposes a substantial increase in the required precision of the calibration process for it to be a meaningful, and not just a perfunctory, cosmetic process. Many data analysis schemes that have appeared in the literature are very sensitive to the precision of the calibration process; some are much less so. These developments in sensor technology suggest the added value of those of the latter type compared to the former.

Systematic Distortion Removal It is sometimes desirable to suppress types of systematic distortion during the preprocessing process. As an example of this, consider the Landsat MSS scanner. It used six detectors in each of its four spectral bands in the scanning process, scanning out six lines at once, as shown in Figure 10-5.

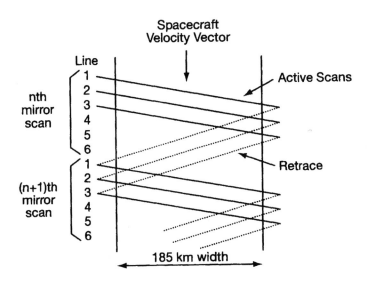

Figure 10-5. The scan pattern for Landsat MSS

453

The sensitivity of the six detectors would undoubtedly vary slightly from one another, as might be expected. Thus, particularly over uniform targets such as bodies of water, a slight stripping will be visible in imagery constructed from the data. It is possible to remove or at least greatly mitigate this effect since it occurs in a quite systematic and predictable way. One way is to adjust the digital count of the six lines so that they provide a uniform response.

As a second example, suppose that it is discovered after the launch of a spacecraft that there is an unanticipated mechanical vibration, electrical crosstalk, or oscillation at a certain frequency that produces an additive, fixed frequency component to the detector output signal. This could be removed or reduced by applying a notch filter in which the notch frequency is adjusted to the offending frequency. Note that in doing so, some signal is also lost, specifically that portion of the signal spectrum that exists at the notch frequency.

Another important kind of distortion can occur in sensors based upon two dimensional array detectors. One dimension of the array senses in the cross-track spatial direction, while the other direction of the array is the wavelength direction. A shift in the geometry of the wavelength deflection mechanism, e.g. a shift in the prism alignment due to the rigors of launch or a changing on-orbit thermal environment, would mean that different wavelengths are being directed to the individual detectors than intended, thus distorting the wavelength calibration. One means that might be used to sense such an error would be to periodically view a fixed wavelength source such as a laser.

10.3 Geometric Preprocessing

Distortion Removal There are many kinds of geometric distortion that can be introduced during the data collection process. Figure

10-6 illustrates some of the more important kinds. The arrow marked S/C indicates the direction of spacecraft travel during the data collection period.

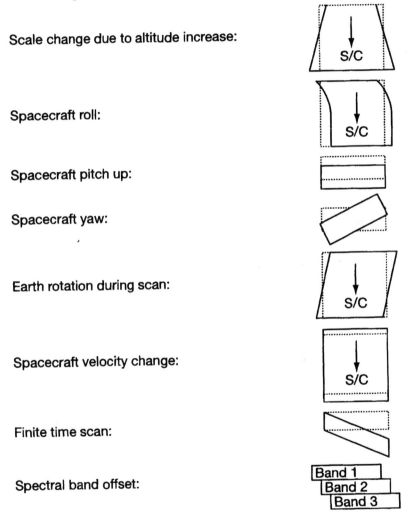

Scale change due to altitude increase:

Spacecraft roll:

Spacecraft pitch up:

Spacecraft yaw:

Earth rotation during scan:

Spacecraft velocity change:

Finite time scan:

Spectral band offset:

Figure 10-6. Some sources of geometric distortion.

Geometric distortion adjustment involves two basic processes:

1. Determining the **proper geometric locations** for the points on a new grid of pixels, and

2. Determining the **proper radiometric values** to use for each of these new grid points.

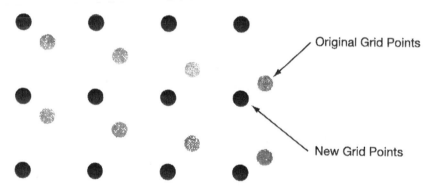

Figure 10-7. Original (distorted) and adjusted image grid patterns.

Geometric location can be obtained either by so-called model-based or parametric means or by nonparametric means. In the model-based approach the distortion is represented using functions specifically describing the physical effects, e.g., the measured yaw angle or the estimated Earth's rotation during the data collection interval. In the non-parametric approach, general polynomials are used to represent the aggregate of a number of physical effects without attempting to model or identify each separately. Ground control points or shapes, i.e. points on the ground whose precise location is otherwise known or physical objects such as a known straight road, may be used to provide more precision to the process.

456

In order to gain additional insight into this process, we define two Cartesian coordinate systems, as in Figure 10-8. We will refer to the uv system as the reference domain, meaning the corrected coordinate system, and the xy system as the data domain.

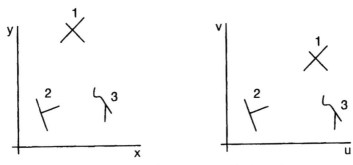

Figure 10-8. Geometric image coordinate systems.

The problem is to map all of the points in the xy data domain to the uv reference domain. Thus a functional relationship between the two is needed, e.g. $u = f(x,y)$ and $v = g(x,y)$. If an explicit form for this functional relationship is not known or cannot be determined, then a polynomial approximation might be assumed, e.g.

$$u = a_o + a_1x + a_2y + a_3xy + a_4x^2 + a_5y^2$$

$$v = b_o + b_1x + b_2y + b_3xy + b_4x^2 + b_5y^2$$

The next step would be to determine the constants $\{a_n, b_n\}$. This can be done through the use of ground control points, points whose correct locations are known in the reference domain and whose location can be determined in the data domain, perhaps by viewing it in image form. Prominent road intersections, stream confluences, and the like are commonly used, as implied by the numbered points in Figure 10-8.

457

The second step in geometric distortion adjustment, determination of the proper radiometric values, is achieved via an interpolation or re-sampling of the image data, that is, by constructing the interpolated pixel values for a new, geometrically correct grid of pixels based on the pixel values of the original geometrically distorted grid of pixels. The problem is illustrated in Figure 10-9.

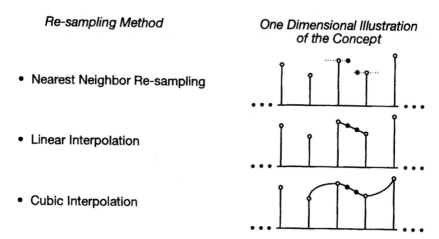

Figure 10-9. One-dimensional illustrations of types of geometric interpolation.

Given the measured radiometric values on the outlined dots representing original pixel centers, what radiometric value should be associated with the new black dot pixels? There are a number of interpolation techniques that are in common use for this re-sampling process. For several of the most common, the sketches above give a one-dimensional illustration of the concept that must be followed in two dimensions.

The extension to two dimensions is straightforward in the nearest-neighbor case. The linear interpolation case above becomes the so-called bilinear transformation in two dimensions. This involves the fitting of three line segments to data in the reference domain, as

458

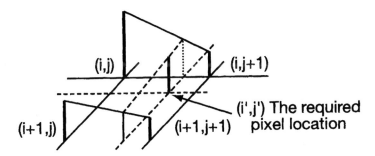

Figure 10-10. Two-dimensional bilinear interpolation.

shown in Figure 10-10, and then interpolating the needed value from the third of these three.

It is seen in Figure 10-11 that cubic interpolation in one dimension involves fitting a cubic curve to four known points in order to permit interpolating values in between the known points. In two dimensions this generalizes to using the surrounding 16 points, and is known as cubic convolution. The process is similar to bilinear interpolation, except that cubics are involved. Cubic polynomials are fitted along the four lines surrounding the desired location, then a fifth cubic

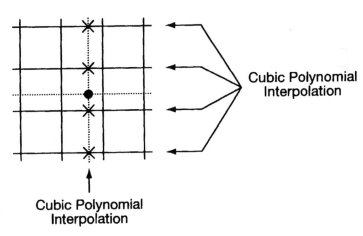

Figure 10-11. Two-dimensional interpolation pattern.

459

polynomial is fitted through these to interpolate to the needed value, as shown in Figure 10-11.

Cubic convolution produces an image that is generally smoother in appearance and is often the interpolation method of choice if the image produced is to undergo analysis by photo interpretation. However, such interpolation does change the radiometric value of each pixel, and thus, if a machine classification algorithm is to be used, the nearest-neighbor method is often preferred.

Geometric Registration A very important type of preprocessing is the geometric registration of data from a scene using two different sources or the same source at two different times. For example, geometric registration is necessary if so-called multitemporal analysis is to be carried out. In multitemporal analysis, the hypothesis is that the manner in which features in the scene change with time provides information as to what the features are. The preprocessing task then is to bring pixels from different data sets of a scene into consonance, as shown in Figure 10-12.

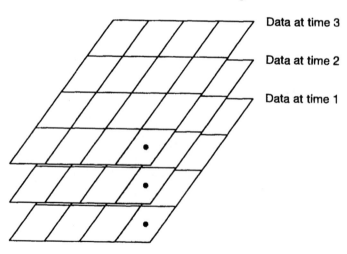

Data at time 3

Data at time 2

Data at time 1

Figure 10-12. Registration of multitemporal image data.

This is in fact a form of geometric adjustment since the need is to make the microdistortions that exist in one data set conform as nearly as possible to those that exist in another.

Again,there are two general approaches to geometric registration:

1. Model-based registration, and

2. Control-point registration

In the model-based approach, the sequence of steps is as follows:

1. The target (Earth) and sensor position, heading, and perturbations are first characterized as a function of time.

2. Systematic distortions due to sensor and sensor/target geometry are adjusted.

3. The computed geographic coordinates are transformed to the desired map projection, as desired, if necessary.

4. All data are registered when translated to the common coordinate system and projection.

The problem with this procedure is that the position and attitude of the platform are usually not known with sufficient precision to obtain registration accuracy to the nearest pixel or better. Thus a model-based adjustment could be used to obtain a first approximation to registration, and then a control-point registration would be used to obtain the final result.

In the control-point approach, the steps are,

461

1. Matching control points are identified in the reference and overlay images.

2. An image mapping function is determined that will register the control points

3. Re-sampling of the overlay image is performed to determine new points in the transformed overlay image.

4. Evaluation of the registration is performed in terms of the residual errors.

Once more, there are at least two approaches to location of control points, manual and calculated. In some cases, the manual approach is used to get a good approximation to the correct locations, followed by use of the calculated method to refine them.

In the manual case, data of the two data sets are viewed in image form to locate corresponding points in the two that are readily identifiable. In the calculated approach, two-dimensional correlation may be used, as follows: Define one image as the reference image, the other being referred to as the overlay image based on the intent that the geometry of the overlay image is to be adjusted to conform as closely as possible to that of the reference image as shown in Figure 10-13. Choose a rectangular window size, N x N pixels, (e.g. N = 20) for correlation purposes. The following calculations are then performed:

Reference image: X_{ij} $i = 1,2,...,N$ $j = 1,2,...,N$
Overlay image: Y_{ij} $i = 1,2,...,N$ $j = 1,2,...,N$

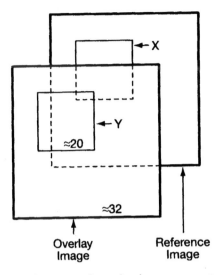

Figure 10-13. The reference and overlay image geometry in image registration.

Window means:

$$m_x = \frac{1}{N^2} \sum_{i=1}^{N} \sum_{j=1}^{N} \mathbf{x}_{ij}$$

$$m_y = \frac{1}{N^2} \sum_{i=1}^{N} \sum_{j=1}^{N} \mathbf{y}_{ij}$$

Correlation of the windows:

$$\rho_{xy} = \frac{\sum_{i=1}^{N} \sum_{j=1}^{N} (\mathbf{x}_{ij} - \boldsymbol{\mu}_x)(\mathbf{y}_{ij} - \boldsymbol{\mu}_y)}{\sqrt{\sum_{i=1}^{N} \sum_{j=1}^{N} (\mathbf{x}_{ij} - \boldsymbol{\mu}_x)^2 \sum_{i=1}^{N} \sum_{j=1}^{N} (\mathbf{y}_{ij} - \boldsymbol{\mu}_y)^2}}$$

The window is moved until $|\rho_{xy}|$ is a maximum. A value of $|\rho_{xy}| \geq 0.5$ is considered a good correlation. Frequently, correlation between two images gathered at different times is not high due to differences that take place in the scene, the very changes that may be information bearing. Thus, for registration purposes, chances of good correlation may be improved by first carrying out an edge enhancement of the two images, and then correlating the edge enhanced data. Another technique used is to first divide the image data into objects, and then determine the centroid of the objects. These centroids are then to be matched between the reference and overlay images.

Figure 10-14 presents some results of a test of the use of multitemporal information and its effect in improving classifier performance. These results indicate first that some times of the season provide better conditions for discriminating among certain

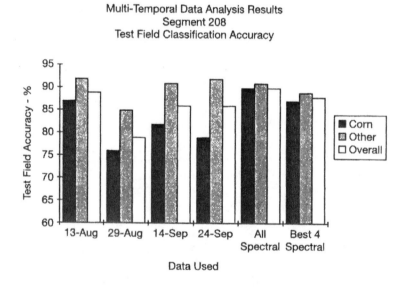

Figure 10-14. Example results for a multitemporal classification problem.

classes than others. They also show that combining observation times can lead to improvement over any specific time.

Figure 10-15 presents some results that assist in determining the accuracy of registration data needed.[118] Thematic Mapper spectral bands and spatial resolution were simulated using higher resolution aircraft data. Then one of the spectral bands used in a test classification was purposely misregistered by successive amounts to determine the effect on the classification. The vertical axis shows the percentage of the pixels whose classification was changed as a

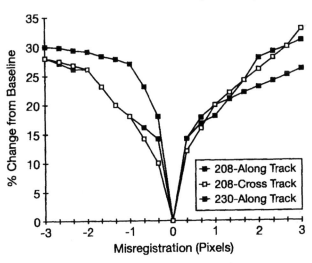

Figure 10-15 (repeat of Figure 5-5). Registration sensitivity in a classification problem.

118 P. H. Swain, V. C. Vanderbilt, and C. D.Jobusch, A quantitative applications-oriented evaluation of Thematic Mapper design specifications, *IEEE Transactions on Geoscience and Remote Sensing*, Vol. GE-20, No. 3, Jul. 1982.

result of the misregistration. Notice that the relationship between misregistration and pixel classification is quite sharp, suggesting the need for subpixel accuracy.

10.4 Goniometric Effects

Attention has often been drawn to the fact that the reflectance of Earth surface materials is especially dependent on the angle of illumination and the angle of view. This fact has a pronounced effect upon all types of remotely sensed data. This effect on imagery is illustrated in Figure 10-16. This image, an air photograph, was gathered at a time prior to local noon. The image was gathered from an aircraft having a northern heading such that at that hour of

Figure 10-16. Air photo of Northwest Indiana prior to local noon, from a north-bound aircraft.

the morning the sun was to the right and somewhat to the rear (due to the season and latitude of the flightline) of the aircraft. The bright area or "hot spot" in the upper left quadrant is not an artifact of image reproduction, but a sun and view angle effect.

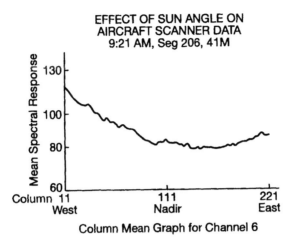

EFFECT OF SUN ANGLE ON
AIRCRAFT SCANNER DATA
9:21 AM, Seg 206, 41M

Column Mean Graph for Channel 6

Figure 10-17. Average spectral reflectance vs. view angle for an example data
set (3 hours before local noon).

The reason for this effectr is that since the line of view is nearly
parallel to the illuminating rays of the sun on the upper left portion
of the image, all surfaces viewed are fully illuminated. In other
words there are no shadowed areas viewed in this portion of the
image. Thus the amount of energy reflected from any given (rough
surface) material tends to be greater. Areas where the view angle
departs from parallel to the sun's rays thus necessarily appear
successively darker. To visualize what the effect would be like from
a line scanned device, focus on a line across the middle of the image.
There should be a bright area on the left side of the flight line relative
to the right. Figure 10-17 shows an example of this effect from line
scanner data.

Plotted in Figure 10-17 is the average response for a large number
of columns of data in the imagery plotted versus the column number.
For this data set, gathered shortly after 9:00 a.m., the response is
clearly greater on the left, or west, portion of the field of view than
it is on the right, or east. Figures 10-18 and 10-19 show the same
type for data gathered near local noon and late in the afternoon,

467

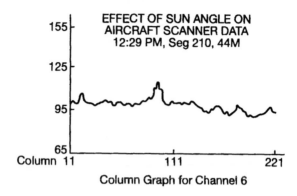

Figure 10-18. Average spectral reflectance vs. view angle for an example data set (local noon).

respectively. Variations of this effect with time of day are readily apparent.

It would be relatively easy to improve the appearance of the imagery by appropriate processing. One could use a curve such as those of these three figures to derive appropriate multiplicative adjustment

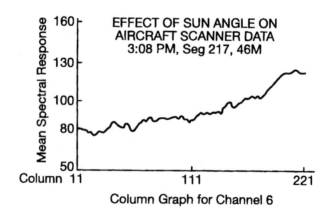

Figure 10-19. Average spectral reflectance vs. view angle for an example data set (2.5 hours after local noon).

factors for each column in the data. Prior to applying the adjustment, this characteristic curve should be smoothed appropriately so that the random variations present in the characteristic curve do no show up in the processed image.Perhaps more important than the appearance of the image, however, is the quantitative quality of the data. A more quantitative test of the relative effectiveness of this type of data adjustment can be obtained by carrying out a

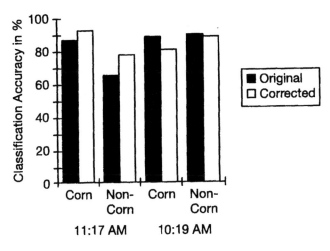

Figure 10-20. Test of accuracy for a simple data correction procedure.

classification of the data set into appropriate classes for the case of both corrected and uncorrected data. Two examples of the result of this classification are shown in Figure 10-20.

As the figure shows, the results for corn versus non-corn classifications were carried out on two different data sets are shown, and in each case the classification of the original versus sun-angle-corrected data is compared. Note that the adjustment of the data did result in a significant improvement in accuracy for the classification indicated by the two pairs of bars on the left. However, the second classification shows a degradation of performance due to data correction. This degradation occurred in the data set collected earlier in the morning, when the sun angle effect is more severe.

The point this illustrates is that though satisfactory improvement in image appearance is possible, the results with regard to improving data quality are quite mixed. The reason for this is that the degree to which the illustrated sun angle effect takes place depends not only on the angle relationship of the sun, the scene, and the observer, but also on the contents of the scene. Individual areas within the scene will display this effect to a greater or lesser extent depending on their contents. This fact is further supported by examining the next few photographs taken after that of Figure 10-16. As the aircraft moves northward, the "hot spot" in the image does as well, as shown in Figure 10-21.

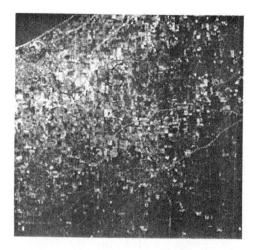

Figure 10-21. Air photo over land showing the "hot spot."

Figure 10-22. Air photo showing the movement of the "hot spot" as the aircraft flies.

Figure 10-23. Air photo showing the "hot spot" vanishing as the aircraft moves over water.

The hot spot continues to be present until the position of the aircraft taking the image occurs over the waters of Lake Michigan, as seen in Figures 10-22 and 10-23.

Since at that point the hot spot is no longer present, this demonstrates the fact that the hot spot was due not only to the sun-angle–view-angle relationship but to surface roughness in the scene. As soon as the illuminated surface-shadowed surface relationship is changed, by the aircraft's movement from land to water, the degree of the sun angle effect is changed materially. The substantial dependence of the magnitude of the effect on the surface roughness means that to preserve the information content of the data, a locally appropriate adjustment of the data would be required, but the information needed for this is usually not available.

Appendix An Outline of Probability Theory

Probability theory has a fundamental role in the field of pattern recognition. Although it is not necessary to develop a thorough background in probability theory before learning to apply pattern recognition techniques to remotely sensed data, a brief outline of the basic concepts may prove useful. This appendix provides such an outline.

Probability theory ordinarily proceeds from a series of axioms and definitions. Based upon these definitions a number of useful relationships can be established with which to solve problems that involve a degree of uncertainty. We will proceed to state these definitions and briefly describe the needed relationships, omitting formal proofs or derivations for brevity. The reader may wish to refer to elementary textbooks on the subject if greater detail is desired.[119]

Random Experiment: An experiment whose outcome cannot be predicted.

Trial: A single performance of a random experiment.

Outcome, α: The result of a trial of the experiment.

Universal Set or Space, S: The entire set of all possible outcomes of the experiment.

---

[119] G. R. Cooper and C. D. McGillem, Probabilistic methods of signal and system analysis, 3rd ed., Oxford University Press, 1999.

Event: A collection of outcomes of particular interest.

Probability: A number associated with an event. The number is (usually) a measure of how likely the event is, and it is usually assumed, although we often think of it conceptually as assigned by its relative frequency of its occurrence. That is, if the experiment were to be performed N times and event A occurred N_A times, then $N_A/N \rightarrow Pr(A)$ as $N \rightarrow \infty$

Here are some of the properties:

- $0 \le P_r(A) \le 1$.
- If a total of only M mutually exclusive events are possible then $P_r(A) + P_r(B) + ... P_r(M) = 1$.
- An impossible event is represented by $P_r(A) = 0$, but if an event has probability $= 0$ that does not mean it is impossible.
- A certain event is represented by $P_r(A) = 1$, but if an event has probability $= 1$ that does not mean it is certain to occur.

Joint Probability: Pr(A,B) is the probability associated with the occurrence of both A and B.

Conditional Probability: Pr(A | B) = probability associated with A given that B is known to have occurred.

Note that $P_r(A,B) = Pr(A \mid B) P_r(B) = P_r(B \mid A) P_r(A)$ and that

$$Pr(A|B) = \frac{Pr(A,B)}{Pr(B)} \quad \text{assuming } Pr(B) > 0$$

Statistical Independence For two events statistical independence is defined by the relation $P_r(A,B) = P_r(A) P_r(B)$

Total Probability

Let $A_1 + A + ... A_n = S$ meaning that $\{A_i\}$ are an exhaustive set of events, and $A_i A_j = 0$, $i \ne j$ meaning that $\{A_i\}$ are a mutually exclusive set. Let M be another event. Then, $Pr(M) = Pr(M|A_1) Pr(A_1) + Pr(M|A_2) Pr(A_2) + \ldots + Pr(M|A_n) Pr(A_n)$

Bayes Theorem

$$Pr(A_i|M) = \frac{Pr(M|A_i)Pr(A_i)}{Pr(M)}$$

$Pr(A_i|M) =$

$$\frac{Pr(M|A_i)Pr(A_i)}{Pr(M \mid A_1) Pr(A_1) + Pr(M \mid A_2) Pr(A_2) + \ldots + Pr(M \mid A_n) Pr(A_n)}$$

Random Variable A random variable is a real valued function defined over the sample space. That is, for a sample space $S = \{\alpha_n\}$, the random variable $X = X(\alpha_i)$ would map each of the outcomes, α_i, to a real number. This is a useful concept for making calculations such as averages with regard to the outcomes.

Probability Distribution Function The probability distribution function is defined as
$$P_X(x) = Pr(X \le x)$$

That is, the probability distribution function of the random variable X is the probability that the random variable X takes on some value less than or equal to x. Since the distribution function is a probability, it must satisfy the properties previously discussed and thus has the following properties:

$$0 \le P_X(x) \le 1 \; -\infty < x < \infty$$
$$P_X(-\infty) = 0, \qquad P_X(\infty) = 1$$
$$P_X(x) \text{ is a non-decreasing function of } x$$
$$Pr(x_1 < X \le x_2) = P_X(x_2) - P_X(x_1)$$

At this point, we may summarize as follows:

Random experiment with
Outcomes: $\alpha_1, \alpha_2, \ldots \alpha_n \rightarrow$ Random variable $X = X(\alpha_i)$
Events $\qquad\qquad\rightarrow X \leq x$, a "standardized" event
Probability $\qquad\qquad\rightarrow \Pr(X \leq x) = P_X(x)$

Probability Density Function The Probability Density Function is defined based upon the Probability Distribution Function.

Definition: $p_X(x) = \dfrac{dP_X(x)}{dx}$

Significance: $\qquad p_x(x)\, dx = \Pr(x < X \leq x + dx)$

Properties: $\qquad p_x(x) \geq 0$

$$\int_{-\infty}^{\infty} p_x(x)dx = 1$$

$$P_x(x) = \int_{-\infty}^{x} p_x(u)du$$

$$\int_{x_1}^{x_2} p_x(x)dx = \Pr(x_1 < X \leq x_2)$$

Note that while for the probability distribution function, the value of the function is probability, for the probability density function, the area under the function is probability. The value of a probability density function at some specific value of x is referred to as a likelihood. It shows the likelihood of that value of x relative to the other values of x.

Expected Value or *Mean Value* The Expected Value or Mean Value of a random variable is defined by the following expression.

$$\bar{x} = E[x] = \int_{-\infty}^{\infty} x p_X(x) dx$$

From this, it can be shown that for any suitably defined function g(x)

$$E[g(x)] = \int_{-\infty}^{\infty} g(x) p_X(x) dx$$

Joint Probability Density Function

For two random variables we have similar definitions. Thus the definition of a joint probability distribution function is

$$P(x,y) = \Pr[X \leq x, Y \leq y]$$

the properties are as follows:

1. $0 \leq P(x,y) \leq 1$ $-\infty < x < \infty$ $-\infty < y < \infty$

2. $P(-\infty, y) = P(x, -\infty) = P(-\infty, -\infty) = 0$

3. $P(\infty,\infty) = 1$

4. $P(x, y)$ is a nondecreasing function as either x or y or both increase.

5. $P(\infty, y) = P_Y(y)$ $P(x,\infty) = P_X(x)$

Joint Probability Density Function

the definition of joint probability density function is

$$p(x,y) = \frac{\partial^2 P(x,y)}{\partial x \partial y}$$

In this case, $p(x, y)\, dx\, dy = \Pr[x < X \le x+dx,\ y < Y \le y+dy]$

Properties:

1. $p(x,y) \ge 0$ $\qquad -\infty < x < \infty \qquad -\infty < y < \infty$

2. $\displaystyle \int_{-\infty}^{\infty}\int_{-\infty}^{\infty} f(x,y)\ dx\ dy = 1$

3. $\displaystyle P(x,y) = \int_{-\infty}^{x}\int_{-\infty}^{y} p(u,v)\,du\,dv$

4. $\displaystyle p_X(x) = \int_{-\infty}^{\infty} p(x,y)\,dy \qquad p_Y(y) = \int_{-\infty}^{\infty} p(x,y)\,dx$

5. $\displaystyle \Pr[x_1 < x \le x_2,\ y_1 < y \le y_2] = \int_{x_1}^{x_2}\int_{y_1}^{y_2} p(x,y)\,dy\,dx$

Joint Expectation The joint expectation of a function of two vaiables is defined as

$$E[g(x,y)] = \int_{-\infty}^{\infty}\int_{-\infty}^{\infty} g(x,y)p(x,y)\,dx\,dy$$

Correlation The correlation between two variables is defined as

$$E[XY] = \int_{-\infty}^{\infty}\int_{-\infty}^{\infty} xy\,p(x,y)\,dx\,dy$$

The concepts of distribution function, density function, and expectation are thus easily extended to more than one dimension.

Random Process or *Stochastic Process* The random process or stochastic process is defined based on the preceding probability theory concepts. It is the means by which more complex systems and signals, ones too complex to be adequately modeled in a deterministic fashion, can be modeled rigorously so that system designs can be based on it. The spectral response of any given Earth surface cover class is such a complex signal. Define the following notation:

- Denote a sample function as x(t). It is the result of carrying out a random experiment one time. It is a random variable. Although it is traditional to use time t as the independent variable, any independent variable may be used. For example, in the multispectral remote sensing case, wavelength λ is the independent variable of interest. The spectral response of a single pixel, then, is such a sample function.

- An Ensemble {x(t)} then is the collection of sample functions associated with the random experiment. The response of a collection of pixels constitutes such an ensemble.

- A random process or stochastic process is defined as the ensemble with its probabilistic description. Thus the collection of responses of a class, or of a whole scene, with their probabilistic description may be spoken of as a random process.

- The value of a sample function $x(t)$ at any one time t_1 defines a random variable denoted as $X(t_1)$ or simply X_1.

- The probabilistic description consists of the joint density function $p_X(x_1, x_2,...)$ of such random variables defined at all times (wavelengths). However, it is generally adequate to use only the joint density function of random variables defined at two times (wavelengths) arbitrarily located.

Auto Correlation Function The correlation between to random variables was previously defined as $E[x_1 x_2]$.

Let $\quad x_1 = x(t_1)$ and $x_2 = x(t_2)$.

The **Autocorrelation Function** of the process is defined as,

$$R_x(t_1, t_2) = R_x(\tau) = \int_{-\infty}^{\infty}\int_{-\infty}^{\infty} x_1 x_2 p(x_1, x_2)dx_1 dx_2$$

For random processes that are stationary (the statistics do not depend on the choice of time origin), the autocorrelation function depends only on one variable $\tau = t_2 - t_1$. Thus,

$$R_x(t_1, t_2) = R_x(\tau) = E[x(t_1)x(t_1+\tau)]$$

Many very practical signal design and analysis problems can be based upon $R_x(\tau)$ for an ensemble of functions. It is a quantity, that on the one hand, characterizes how a signal set behaves, and on the other hand, provides a quite measurable quantity in practical circumstances. It is one of the key quantities that leads to maximum likelihood detection, as is used in classifiers.

Summary

These major concepts and relations of probability theory are needed in establishing the rules and procedures for analysis of multispectral data. They form the foundation by which spectral responses for classes of interest may be modeled quantitatively, and thus can be used to design analysis procedures that will provide optimal performance

Exercises

Chapter 1

1-1. Shown in Figure P1-1 are images made from two bands of multispectral scanner data.

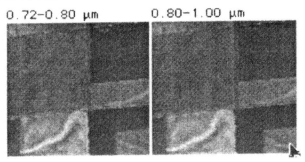

Figure P1-1.

Using sketches of data distributions in 2-dimensional space, present a compelling argument as to what information useful in discriminating between classes may be available in such multivariant data that cannot be discerned from viewing the data in the form of the above imagery.

Chapter 2

2-1. A horizontally oriented object that does not fill the field of view is three meters in length. How long does the object appear to be when viewed from 30° away from vertical?

2-2. An opaque gray body has a reflectance of 0.45. What is its absorption?

2-3. Rayleigh scattering in the atmosphere is the principal reason why on a cloudless day the sky appears to be blue. True or False.

2-4. At a certain wavelength, the Earth at a certain point has a spectral radiant exitance of 50 watts/(m²-μm). At that wavelength the atmosphere has a transmittance of 0.8 and an upward spectral radiant exitance of 10 watts/(m²-μm). What spectral radiant exitance would be measured by a satellite sensor passing over that point?

2-5. Uncalibrated Landsat MSS data over undeveloped land is "clustered", producing three spectral classes having the mean values tabulated below.

		— — — —Means— — — —	
Band	Cluster1	Cluster 2	Cluster 3
1 (.5 -.6 μm)	70	70	100
2 (.6 -.7 μm)	36	36	70
3 (.7 -.8 μm)	20	110	80
4 (.8 -1.1 μm)	5	60	42

Determine which of these clusters is most likely soil, which is most likely water, and which is most likely green vegetation. Show your reasoning clearly.

2-6. Match the following terms with their corresponding values for emittance, ε, by placing the appropriate letter in the blank before the term:

___ Perfect reflector	A.	$\varepsilon = 1$
___ Blackbody	B.	$0 < \varepsilon < 1$
___ Gray body	C.	$\varepsilon = 0$
___ All others	D.	$\varepsilon = f(\lambda)$

2-7. Match the following three means for characterizing the reflectance characteristics of a scene with a property of each by placing the appropriate letter in the space provided.

___ 1. Bi-directional Reflection Distribution Function
___ 2. Bi-directional Reflectance
___ 3. Bi-directional Reflectance Factor

A. Practical but dependent upon the measurement instrument.
B. The most complete description of a surface but impractically
   complex.
C. Practical and not dependent upon the measurement instrument.

2-8. An aircraft carrying a multispectral sensor is flying at an altitude
of 2000 meters above the terrain. The sensor has an IFOV of 2
milliradians and an FOV of 30°. The aircraft is flying at a speed of
75 meters/sec and the sensor senses a new scanline of data every
100 milliseconds.
   a) Find the diameter at nadir of a pixel on the ground in meters.

   b) Find the width of the area on the ground scanned in meters.

   c) What is the largest cross-track diameter of any pixel?

   d) Find the distance in the down-track direction between the
      centers of pixels.

2-9. A spaceborne multispectral scanner is constructed using a
vibrating mirror in a manner similar to that of the Landsat MSS,
but with the following specifications:
   Orbital altitude = 800 km   FOV = 5°
   Orbital rate = 15 km/sec   IFOV = 0.001°
   Determine the following:
   (a) the number of pixels per scan line.
   (b) the size of pixels at nadir.

(c) The time interval for the mirror to scan a single scan line, assuming contiguous scan lines.

2-10. Plot a graph of pixel size in equivalent ground meters as a function of angle from nadir across a swath for the following:
   a) Landsat MSS with IFOV of 0.086 mrad, FOV of 11.56°, H = 920 km.
   b) An aircraft scanner with IFOV of 3.0 mrad, FOV of 80°, H = 1000 m.

2-11. A blacksmith lays a black iron bar on a bed of hot charcoal and starts pumping the bellows to heat the iron. Describe and explain what you will see as the bar gets hotter and hotter. How could you measure the temperature of the bar? What would be the maximum spectral radiant exitance of the bar when it reached 700 °K? (Assume the bar has an emissivity of 0.75.)

2-12. An aircraft flying at 10,000 m about the Earth is carrying a 35mm camera and a spectrometer.
(a) The 35mm camera has a lens with focal length of 135 mm. The field stop in the camera produces a rectangular image that is 35 mm x 24 mm. Sketch and label the dimensions of the ground area that is imaged at any instant. (Assume the camera lens system behaves like a "thin lens."
(b) The spectrometer has a focal length of 50 cm and circular detector with diameter 1.0 mm. What is the angular IFOV of the spectrometer, and what is the size of the ground IFOV?

2-13. A field spectrometer with specifications:
   IFOV = 3.0 mrad
   aperture diameter = 2.0 cm
   wavelength band = 1.0–1.1 μm
is used to make measurements under the following conditions:

486

height above target = 6.0 m

solar zenith angle = 30°

solar spectral irradiance = 700 W/m2-μm over the wavelength band

(a) The instrument views a perfectly diffuse reference standard that fills the field of view and has BRF (bidirectional reflectance factor) of 0.90. The view angle is normal to the surface of the reference standard. What is the spectral radiance in the 1.0–1.1μm band?

(b) While viewing this standard, what is the power incident on the detector over the 1.0-1.1 μm band?

2-14. An airborne multispectral scanner has the following operating characteristics:

IFOV = 3.0 mrad	FOV = 1.5 rad
D = 15 cm	f = 45 cm
q = 1 detectors per band	n = 0.8 (optical efficiency)
p = 2 (2-sided scan mirror)	D/f = 0.33 (optical speed)

v = 500 m/s (aircraft speed during data collection)

scan rate continuously variable from 50 – 250 rev/s

(a) At what altitude should the aircraft fly to obtain a 3.0 m ground resolution?

(b) At what scan rate should the scanner operate to provide contiguous coverage of the ground? (Give the answer in rad/s or rev/s.)

(c) One of the scanner's spectral bands is 0.62–0.68 μm. The targets of interest have a mean spectral radiance of 10 W/m²-μm-sr. The atmospheric transmittance is 0.7. When a target of interest is being viewed, what is the power incident on the 0.62–0.68 μm detector?

(d) For the same targets, it is desired to measure 2% reflectance differences in the 0.62–0.68 μm band. The silicon detector has a detectivity $D = 7.5 \times 10^{11}$ cm $(Hz)^{1/2}$ /W and noise equivalent power NEP = 1.1 x 10^{-11} W. What must be the signal-to-noise ratio for this spectral band?

2-15. A vibrating-mirror multispectral scanner and its spacecraft have the following parameters:

IFOV (ground) = 80 m   f = 82.6 cm
FOV (ground) = 185 km   q = 6 detectors/band
H = 705 km   4 = spectral bands
V = 6.76 km/s   spectral bandwidth = 0.1 μm
D = 22.9 cm   optical efficiency = 0.7
One-directional active scan

Compute the following parameters:
(a) Angular IFOV
(b) Effective detector diameter (assume circular detectors)
(c) Mirror oscillation frequency required for contiguous scanning.
(d) The scanner data rate in bits per second during the active scan if each value is measured to 8-bit precision.
(e) An alternative orbit is being considered with H = 915 km and V = 6.47 km/s. The same ground resolution is desired, so the scanner is to be modified changing only the angular IFOV, the mirror oscillation frequency, and the detector size. Assuming all other parameters remain unchanged, compare the signal-to-noise ratio for the scanner in the alternative orbit to the SNR at the original orbit.

Chapter 3

3-1. Consider the following decision rule:
Decide X belongs to class ω_i if and only if,
$P(X|\omega_i) \geq P(X|\omega_j)$ for all j = 1,2,..., m

Indicate which of the following statements are TRUE or FALSE:

_____ The parameter m is the dimensionality of the measurement vector X.

_____ The decision rule is called the "maximum posterior probability decision rule."

_____ This decision rule is best to use if the prior probabilities of the classes are known.

_____ This decision rule cannot be used if the class distributions are multimodal.

_____ This decision rule cannot be implemented as a nonparametric classifier.

_____ The decision rule is equivalent to the Bayes's decision rule when (1) the prior probabilities are equal, and (2) a 0-1 loss function is used in the Bayes' decision rule.

3-2. A three dimensional measurement

$$X = \begin{bmatrix} 4 \\ 4 \\ 1 \end{bmatrix}$$

is to be classified by means of the following set of discriminant functions:

$$g_1(X) = x_1 + 0.5\, x_2 + 2x_3 + 1$$
$$g_2(X) = 2x_1 + x_2 + 0.5\, x_3$$
$$g_3(X) = 3x_1 + 2x_2 - 4x_3 - 3$$

To which class does X belong?

3-3. The discriminant functions for a 2-class problem in 2-dimensional space are given by

$$g_1(x) = x_1 + x_2 + 2$$

489

$$g_2(x) = 2x_1 + 2x_2$$

a) Write the equation for the decision boundary between the two classes.

b) For a pixel $x = [2, 2]^T$ find the correct class.

3-4. Figure P3-4 contains training data for two classes. Assume a linear decision boundary on the graph for classifying these data with 100% accuracy for a line that passes through the origin and the point $x_1 = 3$ and $x_2 = 4$. Write the equation for the decision boundary and a suitable pair of discriminant functions.

Data for two classes

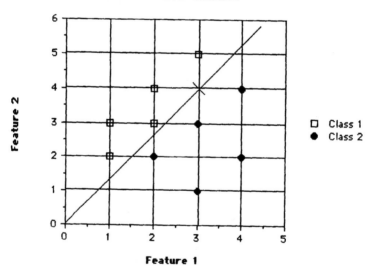

Figure P3-4.

3-5. Using the data given in the table below, where each pixel has two spectral components: (a) Plot the data in a two-dimensional vector space. (b) Compute the sample means and sample covariance

490

matrices for each class and plot the means in the same vector space with the data. (c) Assuming each class to have a two-dimensional normal distribution, compute the values of the class probabilities for each of the three vectors x_1, x_2, x_3, and use the results to classify each pixel vector.

Class 1		Class 2		Class 3	
Band 1	Band 2	Band 1	Band 2	Band 1	Band 2
16	13	8	8	19	6
18	13	9	7	19	3
20	13	6	7	17	8
11	12	8	6	17	1
17	12	5	5	16	4
8	11	7	5	14	5
14	11	4	4	13	8
10	10	6	3	13	1
4	9	4	2	11	6
7	9	3	2	11	3

$$x_1 = \begin{bmatrix} 5 \\ 9 \end{bmatrix} \qquad x_2 = \begin{bmatrix} 9 \\ 8 \end{bmatrix} \qquad x_3 = \begin{bmatrix} 15 \\ 9 \end{bmatrix}$$

3-6. For the same data as in Problem 3-5: (a) Find, by any means, a set of straight lines or line segments that can be used as decision boundaries separating the classes. (b) Draw these boundaries, and derive a set of equations defining the lines or line segments. (c) Now derive a computational rule (an algorithm), using these lines or line segments, that can be used to classify any pixel vector into one of the three classes and show how the rule is used to classify the vectors x_1, x_2, x_3. Comment on both the computational complexity of this rule and its ability to partition the vector space as compared to the rule used in Problem 3-5.

3-7. Funny dice problem. Two players are engaged in a game involving two pairs of dice, one of which is a "standard" pair (with one to six spots on each face), the other an "augmented" pair with two extra spots on each face (three to eight spots per face). Player 1 selects a pair of dice at random from a bin containing 60% standard dice and 40% augmented dice, and rolls the dice, concealing them from player 2. Player 1 announces (honestly) the sum resulting from the roll. Player 2 then guesses which pair was rolled. If player 2 guesses correctly and the dice were "standard", player two wins a dollar. If player 2 guesses correctly and the dice were "augmented", player 2 wins $3. If player 2 guesses incorrectly, he or she loses a dollar. For each of the three classifiers studied, Maximum Likelihood, Minimum Probability of Error, and Minimum Risk, assume that the dice are "fair" and determine the following:

   a) the correct location for the decision boundary.
   b) the probability of being correct on a given trial, and
      the expected winnings after 100 trials.

3-8. Repeat the previous problem, but this time, approximate the probability density function of the expected sum showing on the face of the dice with a Gaussian curve having mean and variance equal to the mean and variance obtained from the ideal distribution used in Problem 3-7.

3-9. Assume now 30-sided dice with the standard dice having from 1 to 30 spots on a side, and the augmented set have an extra 5 spots (i.e. 6 to 35 spots). Repeat Problems 3-7. From your answers to problems 3-7, it appears you could win some money. Would you rather play with these new dice or the other six-sided ones?

3-10. Two classes have probability density functions as shown in Figure P3-10. Class 1 has a prior probability of occurrence of 0.4.

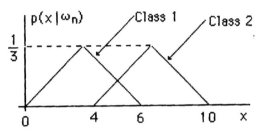

Figure P3-10.

(a) If the decision boundary is set at x= 5, determine the probability of classification error.

(b) Determine the decision boundary location that will produce minimum error.

3-11. A class probability density function is given by

$$p(X \mid \omega_1) = \begin{cases} x/25 & 0 \le x \le 5 \\ 2/5 \pm x/25 & 5 \le x < 10 \end{cases}$$

If a threshold level is set at a likelihood value of 1/25, what portion of the pixels of this class will be thresholded?

3-12. (a) A Gaussian maximum likelihood classifier is trained for two classes, class A with mean = 0 and $\sigma = 1$ and class B with mean = 1 and $\sigma = 1.2$. An unknown sample has a value of 0.6. Determine to which class the unknown will be assigned by the classifier.

(b) If a "minimum distance to means" classifier were used on the classes and the unknown of this problem, to which class would the unknown be assigned.

493

3-13. Match the following types of classifiers with the characteristics in the column at right by placing the appropriate letter next to the classifier type:

_____ Maximum Likelihood

_____ Minimum Probability of Error

_____ Minimum Risk

_____ Parallelepiped

_____ Min. Distance to Means

_____ Fisher Linear Discriminant

_____ Nonparametric

a. makes no assumption about density form

b. accounts for difference in cost of errors by class

c. assumes all classes are govern by the same covariance matrix

d. does not use class variances or covariances.

e. does not use class statistics at all

f. discriminates base upon the estimated class densities alone.

g. requires knowledge of the probability of occurrance of each class

3-14. A given data set contains two classes and can take on the values 1, 2, 3, 4, or 5 with the probabilities shown in Figure P3-14. Class 1 occurs in the data set with probability 0.6.

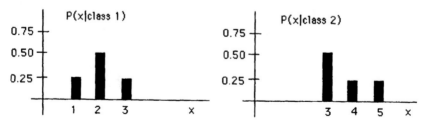

Figure P3-14.

In order to discriminate between them with the minimum probability of error, where should the decision boundary be placed, and what is the expected probability of error.

3-15. A pair of two sided dice designated pair A have one spot on one side and two spots on the other. A second pair of two sided dice designated pair B have two spots on one side and three spots on the other. One of the pairs is randomly selected and rolled, given that the first pair is twice as likely to be selected when the dice are rolled. The total spots appearing is taken as the result. From it you are to estimate which of the two pairs, A or B, was chosen. Determine what your choice would be for each possible outcome using a MAP classification rule and what your expected error rate will be.

Possible outcomes 2   3   4   5   6

Your choice   ____   ____   ____   ____   ____

3-16. Shown below is a table of three channel interclass distances for a certain 5 class classification task.
   (a) Based on this table, which three channel feature set would be expected to provide the highest classification accuracy on the average?
   (b) Which feature set would be expected to provide the greatest accuracy for the least separable class pair of classes?

class pair symbols >			12	13	14	15	23	24	25	34	35	45
weighting factor >			(10)	(10)	(10)	(10)	(10)	(10)	(10)	(10)	(10)	(10)
Chan.	Min.	Ave.	Weighted Interclass Distance Measures									
1 4 5	0.88	3.08	4.61	1.23	0.92	1.85	4.50	0.88	5.87	1.53	6.88	2.54
1 4 6	0.91	2.58	4.58	1.09	0.91	1.70	4.25	0.91	4.25	1.38	4.86	1.84
1 5 6	0.55	2.52	2.77	1.24	0.55	1.21	3.15	0.60	5.35	1.34	6.42	2.58
2 4 5	0.85	3.06	4.50	1.22	0.93	1.84	4.56	0.85	5.71	1.51	6.88	2.55
2 4 6	0.89	2.58	4.41	1.12	0.92	1.74	4.47	0.89	3.93	1.42	5.08	1.83
2 5 6	0.57	2.50	2.74	1.24	0.57	1.21	3.13	0.57	5.14	1.33	6.43	2.60
3 4 5	0.90	3.09	4.53	1.26	0.90	1.89	4.48	0.94	5.79	1.55	6.98	2.55
3 4 6	0.90	2.62	4.49	1.15	0.90	1.75	4.39	0.98	4.21	1.45	5.02	1.84
3 5 6	0.56	2.64	3.08	1.26	0.56	1.27	3.50	0.71	5.41	1.44	6.56	2.59
4 5 6	0.84	2.97	4.28	1.23	0.90	1.93	4.09	0.84	5.43	1.48	7.00	2.55
4 5 7	0.78	2.90	4.18	1.21	0.89	1.82	3.93	0.78	5.47	1.42	6.83	2.47
4 6 7	0.82	2.42	4.16	1.10	0.89	1.69	3.81	0.82	3.83	1.33	4.84	1.78
5 6 7	0.53	2.46	2.71	1.22	0.55	1.23	2.93	0.53	5.07	1.31	6.50	2.54

3-17. (a) For the class-conditional measurement probability density functions given in Figure P3-17, and assuming the class prior probabilities to be $p(\omega_1) = 1/3$ and $p(\omega_2) = 2/3$, find the specific decision rules for (i) the maximum posterior probability, and (ii) the maximum likelihood strategies, i.e. for each case, write the rules in the form,

(i) Decide X in ω_1 if _____   (ii) Decide X in ω_1 if _____
    X in ω_2 if _____   X in ω_2 if _____

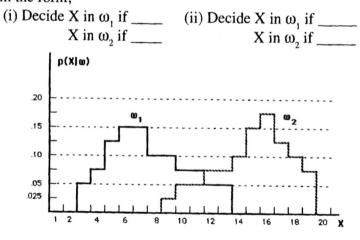

Figure P3-17.

496

(b) Compute the actual probability of error for each of the classification rules in (a).

3-18. A single band data set contains two Gaussian classes of data with mean values of 5 and 2 and σ's of 2 and 1, respectively. From Figure P3-18, determine the expected probability of correct classification for these classes.

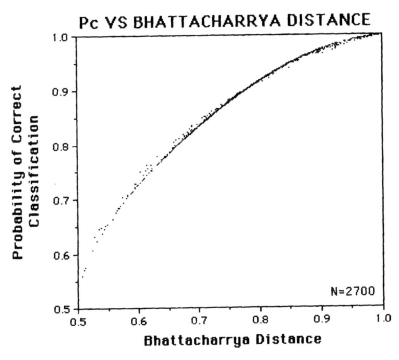

Figure P3-18.

Upon completion of the study of Chapter 3, the Moderate Dimensional Multispectral Data Analysis Project may be begun.

Chapter 4

4-1. The Statistics Enhancement Processor of MultiSpec uses an iterative calculation to maximize the extent to which the equation

$$p(x \mid \theta) = \sum_{i=1}^{m} \alpha_i p_i(x \mid \phi_i)$$

is, indeed, an equality. For this equation, what are the following quantities:

Quantity	Definition
$p(x \mid \theta)$	
m	
α_i	
$p_i(x \mid \phi_i))$	

4-2. In estimating the second-order statistics of classes from a limited number of training samples, it was seen that as the number of features to be used in a given classification increases, the number of parameters that must be estimated increases very rapidly. As a result it was found that it can be that a simplifying assumptions can lead to improved accuracy because the second-order statistics contain fewer parameters that must be estimated with the limited number of training samples. Listed below are several such simplifying assumptions. For a 5-feature, 5-class case, rank them from fewest to most parameters per class that must be estimated by placing a number from 1 (fewest) to 4 (most) in front of them.

Rank   Assumption
_____   Class Covariances
_____   Common Covariance
_____   Diagonal Class Covariances
_____   Diagonal Common Covariance

Chapter 6

6-1. Match the Feature Extraction methods below with the phrases at right by placing the appropriate letters in the blanks.

_____   Principal Components   A. Is based directly on the training samples

_____   Discriminant Analysis   B. Is useful in assisting the image interpreter as to the contents of a pixel

_____   Tasseled Cap   C. Results in features that are statistically uncorrelated.

_____   NDVI   D. Maximizes the between-class to within-class variance ratio.

_____   Decision Boundary   E. Is focused on providing a measure
Feature Extraction   of the relative amount of vegetation in a pixel.

6-2. Give a disadvantage to using a principal components transformation to reduce the dimensionality of hyperspectral data analysis of several classes.

6-3. Compare the advantages and disadvantages of Discriminant Analysis Feature Extraction and Decision Boundary Feature Extraction.

6-4. Justify the statement that with 10-bit hyperspectral data, theoretically any class of material is separable from any other material with vanishingly small error.

6-5. State what the chief limitation is on the statement of Problem 6-4 in any given practical circumstance.

6-6. If two hyperspectral classes have the same mean response at all wavelengths, state how it is possible for the two classes to still be separable.

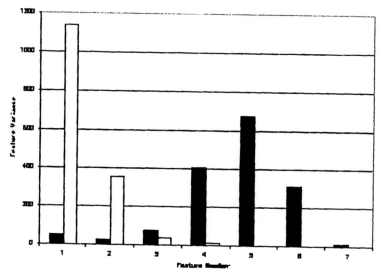

Figure P6-7.

6-7. Shown in Figure P6-7 are bar graphs showing the relative size of the variance of the 7 bands of a Thematic Mapper data set before and after a principal components transformation. Are the black or the white bars those belonging to the principal components transformation? Circle the correct choice.

      Black   White

Upon completion of the study of Chapter 6, the Hyperspectral Data Analysis Project may be undertaken.

Chapter 9

9-1. Match the types of noise in the list below with the phrase best describing a characteristic of that type of noise.

1. Shot Noise	____ Can be calculated as the square root of the sum of the squares of all independent noise sources.
2. Thermal Noise	____ Is related to the bandwidth of the amplifier.
3. Quantization Noise	____ Is related to the level of the signal.
4. Total Noise	____ Depends on the number of bits used to transmit the signal.

Chapter 10

10-1. Shown in Figure P10-1 is a histogram of data from a small area of an image. It is desired to present these image data in an 8 shades of gray form in such a way that each of the 8 gray shades are used as nearly to the same number of times as all of the others. This process is called contrast stretching or histogram equalization. You are to specify to which of the eight new brightness values each of the 32 old brightness values are to be mapped in order to accomplish this by filling in the following table:

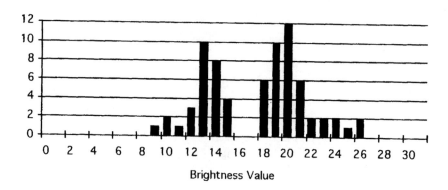

Figure P10-1.

New Brightness Value	0	1	2	3	4	5	6	7
Old Brightness value Range								
Count in new brightness bins								

Index

D

E

F

G

H

I

K

L

M

N

O

P

T

U

V

Printed in the United States of America/BNB

CUSTOMER NOTE: IF THIS BOOK IS ACCOMPANIED BY SOFTWARE, PLEASE READ THE FOLLOWING BEFORE OPENING THE PACKAGE.

This software contains files to help you utilize the models described in the accompanying book. By opening the package, you are agreeing to be bound by the following agreement:

This software product is protected by copyright and all rights are reserved by the author and John Wiley & Sons, Inc. You are licensed to use this software on a single computer. Copying the software to another medium or format for use on a single computer does not violate the U.S. Copyright Law. Copying the software for any other purpose is a violation of the U.S. Copyright Law.

This software product is sold as is without warranty of any kind, either express or implied, including but not limited to the implied warranty of merchantability and fitness for a particular purpose. Neither Wiley nor its dealers or distributors assumes any liability of any alleged or actual damages arising from the use of or the inability to use this software. (Some states do not allow the exclusion of implied warranties, so the exclusion may not apply to you.)

WILEY